Cambridge Elements ≡

Elements in Emerging Theories and Technologies
in Metamaterials
edited by
Tie Jun Cui
Southeast University, China
John B. Pendry
Imperial College London

EFFECTIVE MEDIUM THEORY OF METAMATERIALS AND METASURFACES

Wei Xiang Jiang
Southeast University, China

Zhong Lei Mei
Lanzhou University

Tie Jun Cui
Southeast University, China

MATERIALS RESEARCH SOCIETY®
Advancing materials. Improving the quality of life.

CAMBRIDGE
UNIVERSITY PRESS

MATERIALS RESEARCH SOCIETY®
Advancing materials. Improving the quality of life.

CAMBRIDGE
UNIVERSITY PRESS

University Printing House, Cambridge CB2 8BS, United Kingdom

One Liberty Plaza, 20th Floor, New York, NY 10006, USA

477 Williamstown Road, Port Melbourne, VIC 3207, Australia

314–321, 3rd Floor, Plot 3, Splendor Forum, Jasola District Centre,
New Delhi – 110025, India

103 Penang Road, #05–06/07, Visioncrest Commercial, Singapore 238467

Cambridge University Press is part of the University of Cambridge.

It furthers the University's mission by disseminating knowledge in the pursuit of
education, learning, and research at the highest international levels of excellence.

www.cambridge.org
Information on this title: www.cambridge.org/9781108819183
DOI: 10.1017/9781108872386

First published 2021

A catalogue record for this publication is available from the British Library.

ISBN 978-1-108-81918-3 Paperback
ISSN 2399-7486 (online)
ISSN 2514-3875 (print)

Effective Medium Theory of Metamaterials and Metasurfaces

Elements in Emerging Theories and Technologies in Metamaterials

DOI: 10.1017/9781108872386
First published online: December 2021

Wei Xiang Jiang
Southeast University, China

Zhong Lei Mei
Lanzhou University

Tie Jun Cui
Southeast University, China

Author for correspondence: Wei Xiang Jiang, wxjiang81@seu.edu.cn

Abstract: Metamaterials, including their two-dimensional counterparts, are composed of subwavelength-scale artificial particles. These materials have novel electromagnetic properties and can be artificially tailored for various applications. Based on metamaterials and metasurfaces, many abnormal physical phenomena have been realized, such as negative refraction, invisible cloaking, abnormal reflection, and focusing, and many new functions and devices have been developed. The effective medium theory lays the foundation for design and application of metamaterials and metasurfaces, connecting metamaterials with real-world applications. In this Element, the authors combine these essential ingredients and aim to make this Element an access point to this field. To this end, they review classical theories for dielectric functions, effective medium theory, and effective parameter extraction of metamaterials, also introducing front-edge technologies like metasurfaces with theories, methods, and potential applications. Energy densities also are included.

Keywords: metamaterial, effective parameters, effective medium theory, metasurface, artificial medium

ISBNs: 9781108819183 (PB), 9781108872386 (OC)
ISSNs: 2399-7486 (online), 2514-3875 (print)

Contents

1 Introduction

Artificially structured electromagnetic (EM) materials, also called metamaterials, have received considerable attention over the past 20 years. Metamaterials have exhibited a lot of EM responses not readily found in naturally existing materials [1]. Ever since the left-handed medium with negative permittivity and negative permeability simultaneously was realized and negative refraction was experimentally demonstrated in 2001 [2], metamaterial designs have developed from simple to complicated, and from homogeneous to inhomogeneous. The introduction of inhomogeneous metamaterials with gradient refraction index has resulted in advanced optical lenses and other devices [3, 4], and even invisibility cloaks and optical illusion devices with anisotropic material properties [5–7].

It is conceptually convenient to replace a collection of artificial structures whose elements are in subwavelength scales by a continuous medium, in which the EM properties result from an average of the local EM fields and electric current distributions [8, 9]. In the ideal case, the EM response in the effectively continuous material should be the same as that in the artificial structure. This equivalence can be obtained when the applied fields are static or the element size of metamaterials is much smaller than the free-space wavelength, in which case the artificial structure is called an *effective medium*. In such a case, the inhomogeneous structures can be homogenized from an EM-wave point of view. The procedure of homogenization will enable the effective constitutive parameters (e.g. the electric permittivity and the magnetic permeability) to be defined and used to characterize the artificial structures.

It is very convenient to describe the complex artificial medium by using the bulk EM parameters, which extends the homogenization methods from the traditional regime to modern metamaterial. The retrieved EM parameters are very useful in the design of artificial materials and in the interpretation of their scattering and/or propagation properties. Indeed, many recent activities in EM composites have demonstrated that the scattering elements whose dimensions are an appreciable fraction of wavelength can be described in practice by the effective medium parameters, such as the electric permittivity and magnetic permeability [2–7]. Hence, the use of effective constitutive parameters has been verified successfully in describing and predicting the properties of EM wave propagation and scattering in metamaterials.

As has been pointed out, metamaterials have used man-made atoms to mimic real materials for the flexible controls of EM waves, in which physical principles (e.g. geometrical optics, physical optics, and transformation optics [1, 3, 5–7]) are usually employed to find out the required medium parameters for such controls. Then the question arises of how to design the man-made atoms of metamaterials rapidly and accurately to reach the required medium parameters. This Element seeks to answer this question and aims to establish a bridge between the effective medium parameters of continuous media and the subwavelength meta-atoms of artificial structures through effective medium theory. Sometimes, the rigorous analysis of this equivalence may not be possible; however, the classical model (e.g. Lorentz model) can be used for numerical optimizations. Moreover, a deep understanding of natural materials helps to improve the metamaterial design and create new, controllable metamaterials. Hence, in Section 2, we give a few theories that are usually employed in metamaterials for physical analysis, which lays the theoretical foundation of later chapters. Then in Section 3, we introduce and discuss the retrieval methods of effective medium parameters of artificial structures for quantitative analyses of homogeneous, inhomogeneous, and even bianisotropic metamaterials. In Section 4, we discuss the homogenization of metamaterials by field averaging, which will result in a general effective medium theory of metamaterials, as presented in Section 5, to set up a link between the microscopic meta-atom and the macroscopic effective medium parameters. In the process of deriving the effective medium parameters, we also reveal the EM energy densities in the artificial metamaterials, as demonstrated in Section 6. Finally, we present the effective medium theory, analysis method, and application of metasurfaces for two-dimensional (2D) metamaterials.

2 Classical Theories for the Dielectric Function

When analyzing metamaterials, people usually try to avoid the detailed structure of the material and treat it as a homogeneous one with effective permittivity and/or permeability. Hence, it appears important to have some knowledge about classical theories for dielectric functions. Moreover, in certain cases, the mixing rules for inhomogeneous materials can even lead to the realization of special metamaterials. So, in this Section, we introduce some classical models for the description of natural materials, including the Lorentz model, the Drude model, the Lorentz–Drude model, and the Brendel–Bormann model. Typical mixing equations for inhomogeneous materials are also given, including the Clausius–Mossotti equation, the Maxwell–Garnett equation, and others.

2.1 Lorentz Model

All materials consist of large amounts of atoms or molecules, and thus the interaction between a material and the external EM field should be considered as the collective interactions between these particles and the field. Doing so will

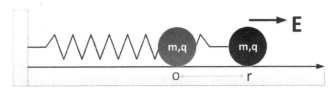

Figure 2.1 The spring model.

definitely involve lots of calculations, if it is not impossible. In this regard, most would be very happy to be able to use a macroscopic parameter, the electric permittivity or magnetic permeability, to simplify this tedious process. Among various theories, the Lorentz model seems to be a very simple and useful tool for the understanding of dielectric functions of various materials, and their variations with the working frequency.

In the Lorentz model, each atom is treated as a harmonic oscillator [10]. The electric force between the nucleus and electron is represented as a spring, which exerts a restoring force on the electron when it leaves the balanced position, as illustrated in Fig. 2.1. We remark that the motion of the nucleus is ignored due to its heavy mass compared with that of the electron.

Suppose that a time-harmonic electric field E is applied, under which the electron will oscillate as illustrated in Fig. 2.1. Using Newton's second law of motion, we have

$$m\ddot{r} = qE - \omega_0^2 mr - \gamma m\dot{r}. \tag{2.1}$$

The first term on the right side of the equation represents the electric force, and the second one indicates the restoring force, which is proportional to the displacement. Here, $\omega_0^2 = k/m$, and k is the stiffness coefficient of the spring. The last term is called the damping force, proportional to the velocity of the particle, in which γ is the damping factor. The dot (or dots) above the r means a derivative with respect to time. Using the Fourier transform, Eq. (2.1) can be easily solved as

$$r = \frac{q}{m} \frac{E}{\omega_0^2 - \omega^2 + i\gamma\omega}. \tag{2.2}$$

When the electron is moved r away from the balanced position, there will be a dipole moment between the nucleus and the electron, which is

$$p = qr = \frac{q^2}{m} \frac{E}{\omega_0^2 - \omega^2 + i\gamma\omega}. \tag{2.3}$$

Suppose that the number density of the atom is N. Then the polarization P, which means the dipole moment per unit volume, should be

$$P = Np = \frac{Nq^2}{m} \frac{E}{\omega_0^2 - \omega^2 + i\gamma\omega}. \tag{2.4}$$

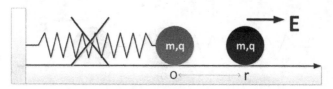

Figure 2.2 The modified spring model.

When the exerted electric field is not large, P is proportional to E (i.e. $P = \varepsilon_0 \chi_e E$). Hence we have

$$\chi_e = \frac{P}{\varepsilon_0 E} = \frac{Nq^2}{\varepsilon_0 m} \frac{1}{\omega_0^2 - \omega^2 + i\gamma\omega}, \tag{2.5}$$

which immediately suggests the dielectric function of the material:

$$\varepsilon_r = 1 + \chi_e = 1 + \frac{\omega_p^2}{\omega_0^2 - \omega^2 + i\gamma\omega}. \tag{2.6}$$

This is the famous Lorentz model for the dielectric function of an arbitrary material [11]. Please note that $\omega_p = \frac{Nq^2}{\varepsilon_0 m}$ is the plasma frequency (angular), which depends on the number density of the particle, the charge of an electron, and the electron's effective mass.

2.2 Drude Model

The Lorentz model works very well for most dielectrics, but not for metals. As we know, metals differ from dielectrics in their microscopic structures, where the valence electrons are not limited by the nuclei and can move freely inside the metal. In many cases, we can treat the metal as an "ocean" of free electrons. Taking this into consideration, the abovementioned Lorentz model can be slightly modified to deal with the difference. As illustrated in Fig. 2.2, we can cut off the spring for the harmonic oscillator to reflect this "free" property [12]. For the deduction process, the only thing we need to do is to set $\omega_0 = 0$. Then, Eq. (2.6) can be reformulated as

$$\varepsilon_r = 1 - \frac{\omega_p^2}{\omega^2 - i\omega\gamma}. \tag{2.7}$$

This is the dielectric function for the Drude model. When the damping factor d is very small, a common case in most metals, Eq. (2.7) can be further simplified as

$$\varepsilon_r = 1 - \frac{\omega_p^2}{\omega^2}. \tag{2.8}$$

Obviously, when the working frequency is smaller than the plasma frequency, the permittivity will be a negative value, which means that the EM waves cannot propagate in the material.

The Drude model can be used to qualitatively explain the property of the ionosphere, the region of the atmosphere between 50 and 500 km above the earth, that can affect radio-wave propagation. Suppose that the electron density inside the ionosphere is about $10^6/cm^3$; this will lead to a plasma frequency at the order of 10^7 Hz. For lower frequencies, the radio waves cannot propagate inside the ionosphere and are reflected back; while for higher frequencies, the ionosphere becomes transparent. This explains why satellite communications must rely on frequencies higher than the maximum plasma frequency of the ionosphere [11].

2.3 Lorentz–Drude Model

Based on the preceding analysis, we observe that the Drude model can be considered as a special case of the Lorentz model, where the oscillating frequency is zero. (We remark that, historically, the two models were derived separately.) The connection between the two models can be further demonstrated if we combine them into one theoretical frame (i.e., the Lorentz–Drude model), which is

$$\varepsilon_r = 1 - \frac{\omega_{pe}^2}{\omega^2 - i\gamma_e\omega} + \frac{\omega_{p1}^2}{\omega_1^2 - \omega^2 + i\omega\gamma_1}, \tag{2.9}$$

in which the subscripts are used to differentiate the parameters. Usually, the first two terms are used to reflect the free-electron effect (intraband effect), and the third one is for the bound-electron effect (interband effect). Moreover, as implicitly suggested in Eq. (2.9), more oscillators can be included in the formula to account for more-complex materials. Hence we have

$$\varepsilon_r = 1 - \frac{\omega_{pe}^2}{\omega^2 - i\gamma_e\omega} + \Sigma_j \frac{f_j\omega_{pj}^2}{\omega_j^2 - \omega^2 + i\omega\gamma_j}, \tag{2.10}$$

where f_j is the strength for the jth oscillator [11].

2.4 Brendel–Bormann Model

Recently, Brendel and Bormann have proposed a model for the dielectric function of solids that replaces the Lorentz oscillator with a superposition of an infinite number of oscillators, given by

$$\chi_j(\omega) = \frac{1}{\sqrt{2\pi}\sigma_j} \int_{-\infty}^{+\infty} \exp\left[-\frac{(x-\omega_j)^2}{2\sigma_j^2}\right] \times \frac{f_j\omega_p^2}{(x^2-\omega^2)+i\omega\Gamma_j} dx. \tag{2.11}$$

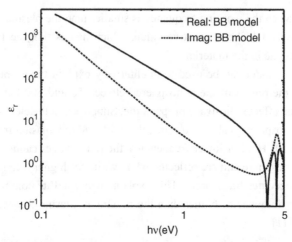

Figure 2.3 The fitted permittivity of Ag using the BB model. The curve has very good agreement with experimental data (not shown in the figure).

Hence,

$$\varepsilon_r(\omega) = 1 - \frac{\omega_p^2}{\omega(\omega - i\Gamma_0)} + \Sigma_j \chi_j(\omega). \tag{2.12}$$

From Eq. (2.11), we see that the jth oscillator is a weighted sum of infinite oscillators, whose resonating frequency is continuous and varies from $-\infty$ to ∞, and the weighted function follows a Gaussian profile, with σ_j being the standard deviation and ω_j its expectation. In this regard, it is clear that each summation term in Eq. (2.10) is a very rough approximation for the continuous form. In other words, the latter is more general than the former. It has been shown that this model can be used to describe optical properties of a wide range of materials including metals [12]. Fig. 2.3 shows the real and imaginary parts of the optical dielectric function of Ag, using the BB model. The curve agrees very well with known experimental data.

2.5 Mixing Rules for Inhomogeneous Materials and Metamaterials

In Sections 2.1 through 2.4, we have shown some classical theories for the dielectric functions of homogeneous materials. In this section, we give some formulas for evaluation of inhomogeneous materials, usually obtained by mixing two or more materials together.

2.5.1 Clausius–Mossotti Equation (or Lorentz–Lorenz Equation)

The Clausius–Mossotti (C–M) equation was first derived to calculate the electric permittivity of a homogeneous material (a macroscopic quantity) using polarizability of its constitutive atoms or molecules (a microscopic quantity). In this regard, it seems inappropriate to put the content in this section. However, this equation is often used to get other mixing rules and usually appears in the same context. For this reason, we list it here for easy understanding and direct comparison of several formulas.

Suppose that there is a uniform electric field E inside a homogeneous material, which is composed of atoms or molecules at the grid point of a cubic lattice. The polarizability of the atom or molecule is α. At first glance, one might expect to express the dipole moment of the particle as $p = \varepsilon_0 \alpha E$. However, this is not the case. As can be imagined, an electric field inside the lattice is not a uniform value. It should change abruptly if one takes a stroll inside the lattice. The field should be very large near the grid point, but change to small values between grid points. What we call a uniform field is actually an averaged quantity inside the material. As a result, one has to find the "local" field acting on a representative particle at the lattice point using the average electric field. Here, we use the method proposed by H. A. Lorentz about 100 years ago [13–17].

Let us arbitrarily choose a grid point inside the lattice structure, say point O, and draw a spherical surface with O as its center. Then the material is naturally divided into two parts, the interior and exterior. Using the principle of superposition, it is quite clear that the electric field at O should be

$$E = E_{i-\text{hom}} + E_{e-\text{hom}}, \tag{2.13}$$

where i represents interior, e represents exterior, and "hom" means homogeneous, indicating that we are treating a continuous and homogeneous material. However, this is not the case, especially when one sits around O and looks around. What one really finds is the discrete lattice structure with polarized particles (atoms or molecules), as shown in Fig. 2.4. Taking this into consideration, Eq. (2.13) should be slightly modified as

$$E_{loc} = E_{i-\text{dis}} + E_{e-\text{hom}}, \tag{2.14}$$

where "dis" means discrete. Subtracting Eqs. (2.13) and (2.14), it is clear that

$$E_{loc} = E + E_{i-\text{dis}} - E_{i-\text{hom}}. \tag{2.15}$$

Due to the symmetry of the lattice structure, E_{i-dis} can be directly evaluated as 0; while for E_{i-hom}, which means the electric field at the center of a sphere due to the homogeneously polarized, spherically shaped bulk material [15], can be easily calculated, giving rise to

$$E_{loc} = E + P/3\varepsilon_0, \tag{2.16}$$

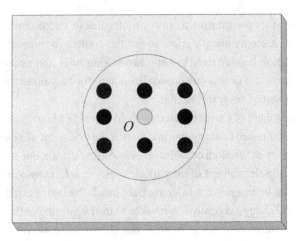

Figure 2.4 Discrete and continuous parts of the material.

where P is the polarization of the material: $P = Np = N\varepsilon_0\alpha(E + P/3\varepsilon_0)$. Since $\mathbf{D} = \varepsilon_0\varepsilon_r\,E = \varepsilon_0 E + P$, we then get the final expression of the dielectric function of the material, which is

$$\varepsilon_r = 1 + \frac{N\alpha}{1 - N\alpha/3}. \tag{2.17}$$

Sometimes, it can also be expressed as

$$\frac{\varepsilon_r - 1}{\varepsilon_r + 2} = \frac{N\alpha}{3}. \tag{2.18}$$

This is the famous Clausius–Mossotti equation. When we substitute $\varepsilon_r = n^2$ into this equation, it becomes what is known as the Lorentz–Lorenz equation. As pointed out by Born et al., Eq. (2.18) was discovered independently nearly at the same time by two scientists of almost identical names, H. A. Lorentz and L. Lorenz [16].

2.5.2 Maxwell–Garnett Equation

Suppose that we have a cubic lattice with spherical and polarizable particles located at each grid point. The permittivity of each particle is ε_m. The distance between the neighboring particles is supposed to be large so that an arbitrarily polarized particle does not greatly influence its neighbors. However, the distance should also be small compared with the working wavelength, a requirement that is certainly satisfied by the static or quasi-static condition.

Due to the abovementioned reasons, we can analyze an arbitrary and isolated sphere, immersed in a uniform electric field. This classical problem can be readily solved without any difficulties [15]. As we know, the polarized sphere acts as a dipole, whose moment p can be expressed as

$$p = 3\varepsilon_0 V \frac{\varepsilon_m - \varepsilon_0}{\varepsilon_m + 2\varepsilon_0} E, \tag{2.19}$$

where V is the volume of the sphere. Eq. (2.19) then indicates that the polarizability of the sphere can be expressed as

$$\alpha = 3V \frac{\varepsilon_m - \varepsilon_0}{\varepsilon_m + 2\varepsilon_0}. \tag{2.20}$$

If we treat the structure as an effectively homogeneous medium, then using the C–M equation, Eq. (2.20) directly leads to

$$\frac{\varepsilon_r - 1}{\varepsilon_r + 2} = f \frac{\varepsilon_{rm} - 1}{\varepsilon_{rm} + 2}, \tag{2.21}$$

where $f = NV$ is the volume ratio of the effective medium. Please note that relative permittivities are used in this equation.

For the spheres embedded in a background medium with relative permittivity ε_{rb}, Eq. (2.21) can be easily extended as

$$\frac{\varepsilon_{reff} - \varepsilon_{rb}}{\varepsilon_{reff} + 2\varepsilon_{rb}} = f \frac{\varepsilon_{rm} - \varepsilon_{rb}}{\varepsilon_{rm} + 2\varepsilon_{rb}}. \tag{2.22}$$

In his famous paper [18] explaining the colors of different glasses with spherical metallic inclusions, Garnett derived the same formula, which reads as

$$\varepsilon_{reff} = 1 + \frac{3f \frac{\varepsilon_{rm} - 1}{\varepsilon_{rm} + 2}}{1 - f \frac{\varepsilon_{rm} - 1}{\varepsilon_{rm} + 2}}. \tag{2.23}$$

Please note that in the original paper, the refractive index, instead of permittivity, is used in the expression. A similar expression holds for the spheres inside a background matrix:

$$\varepsilon_{reff} = \varepsilon_{rb} + \frac{3f \varepsilon_{rb} \frac{\varepsilon_{rm} - \varepsilon_{rb}}{\varepsilon_{rm} + 2\varepsilon_{rb}}}{1 - f \frac{\varepsilon_{rm} - \varepsilon_{rb}}{\varepsilon_{rm} + 2\varepsilon_{rb}}}. \tag{2.24}$$

Equations (2.21)–(2.24) are called Maxwell–Garnett equations.

G. B. Smith gave another derivation for the Maxwell–Garnett equation, which is simple and easily understandable [19]. Suppose that we have a mixture with spherical inclusions with dielectric constant ε_m, and also suppose that the background medium has a dielectric constant ε_b. Then the core of the idea is to focus on one sphere and its surrounding background medium while treating the remaining part as a homogeneous one with an effective permittivity ε_{eff}. (Please refer to Fig. 2.5 for more details). Then, a uniform electric field outside the exterior sphere should not be distorted, since the effective permittivity inside the sphere also constitutes the same effective permittivity.

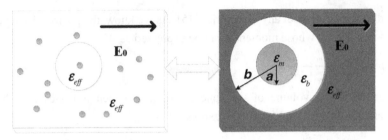

Figure 2.5 Smith's method to derive the M–G equation.

In the right part of the figure, it is quite easy to express the potentials in each region. For example, the potential outside the sphere reads

$$\phi_e = [-E_0 r + D r^{-2}] P_1 (\cos \theta). \tag{2.25}$$

Then, the elimination of the coefficient D, which denotes the field distortion, will give rise to the following result:

$$\begin{aligned} \varepsilon_{reff} &= \frac{\varepsilon_b [b^3 (2\varepsilon_b + \varepsilon_m) + 2a^3 (\varepsilon_m - \varepsilon_b)]}{b^3 (2\varepsilon_b + \varepsilon_m) - a^3 (\varepsilon_m - \varepsilon_b)} \\ &= \frac{\varepsilon_b [(2\varepsilon_b + \varepsilon_m) + 2f(\varepsilon_m - \varepsilon_b)]}{(2\varepsilon_b + \varepsilon_m) - f(\varepsilon_m - \varepsilon_b)}, \end{aligned} \tag{2.26}$$

where $f = a^3 / b^3$ is the volume ratio of the inclusions. Eq. (2.26) is actually the M–G equation.

2.5.3 Bruggeman's Unsymmetrical Formula

As mentioned in Section 2.5.2, for the Maxwell–Garnett formulas, the filling ratio cannot be very large so that the interactions between adjacent particles are weak. This limitation was successfully solved by Bruggeman using an iterative method. In this subsection, we present this process and derive Bruggeman's unsymmetrical formula. Suppose that a material with permittivity ε_m is mixed with a host matrix with permittivity ε_b, and the filling ratio is f_0, which may be large. Then Bruggeman's iterative method will involve mixing the material with the background step by step as follows [20, 21]:

(1) At the initial stage, no material is mixed with the background.
(2) Then, we measure a small quantity of material (say its volume is δV) and mix it with the host matrix (its volume is 1).
(3) The mixture is fully stirred and completely combined, which results in a "new" host medium.
(4) Repeat Steps (2) and (3) until the filling ratio of the material reaches f_0.

This process can be vividly interpreted in Fig. 2.6.

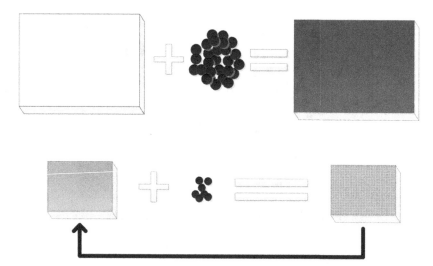

Figure 2.6 Schematic illustration of the mixing process.

Let us look at the mixing process at an arbitrary step when we have a newly mixed host medium with permittivity ε_{eff}. Then we put additional small amounts of material into the host medium, whose volumes are δV and 1, respectively. Since δV is very small, we obtain, according to Eq. (2.24),

$$\delta\varepsilon_{reff} = \frac{3f\varepsilon_{reff}\frac{\varepsilon_{rm}-\varepsilon_{reff}}{\varepsilon_{rm}+2\varepsilon_{reff}}}{1-f\frac{\varepsilon_{rm}-\varepsilon_{reff}}{\varepsilon_{rm}+2\varepsilon_{reff}}} \approx 3f\varepsilon_{reff}\frac{\varepsilon_{rm}-\varepsilon_{reff}}{\varepsilon_{rm}+2\varepsilon_{reff}}, \tag{2.27}$$

where f is the current filling ratio (for the material with volume δV and the current host medium with volume 1). At present, $f = \frac{\delta V}{1+\delta V} \approx \delta V$. Please note that for the mixture of material ε_m and the real host medium ε_b, the actual filling ratio is different, which, if denoted by F before the current-step mixing operation, reads

$$\frac{1*F+\delta V}{1+\delta V} = F + \delta F. \tag{2.28}$$

In Eq. (2.28), the numerator represents volumes of the original material, the denominator represents the total volume of the mixture, and the right side represents the actual filling ratio after the mixing. Then we have

$$\delta V = \frac{\delta F}{1-F}. \tag{2.29}$$

Using Eqs. (2.28) and (2.29), Eq. (2.27) can be rewritten as

$$d\varepsilon_{reff} \approx 3\varepsilon_{reff}\frac{dF}{1-F}\frac{\varepsilon_{rm}-\varepsilon_{reff}}{\varepsilon_{rm}+2\varepsilon_{reff}}. \tag{2.30}$$

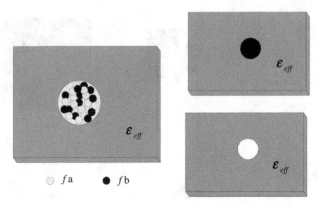

Figure 2.7 Derivation of the mixing formula.

Note that F ranges from 0 to f_0, and the two media are incrementally mixed. Integrating Eq. (2.30) gives rise to

$$\varepsilon_{reff} = \frac{\varepsilon_b}{(1-f_0)^3}(\frac{\varepsilon_{rm} - \varepsilon_{reff}}{\varepsilon_{rm} - \varepsilon_b})^3. \tag{2.31}$$

This is the unsymmetrical Bruggeman's formula. Here, "unsymmetrical" means that the two media are not equivalent in Eq. (2.31), and the interchange of them will yield a different effective value.

2.5.4 Bruggeman's Symmetrical Formula

Suppose that we have a two-component medium, and the filling ratios of the two components are f_a and f_b, respectively, in which $f_a + f_b = 1$. Let us consider an arbitrary part of the medium, as indicated by the circle in Fig. 2.7.

Inside the circle, we have some spherical particles for both a and b. We assume that the effective permittivity of the mixture is ε_{eff}, as illustrated in Fig. 2.7. Then, it is very clear that if the material inside the circle is different from that outside the circle, there will be field deviations when a uniform electric field is applied. However, if the materials are the same, no field deviations occur. Now we focus on the internal part of the circle. If a sphere made of medium a is put inside the effective medium, then, according to Eq. (2.19), the equivalent dipole moment is

$$\boldsymbol{p} = 3\varepsilon_{eff}V\frac{\varepsilon_a - \varepsilon_{eff}}{\varepsilon_a + 2\varepsilon_{eff}}\mathbf{E}. \tag{2.32}$$

Similarly, the result for medium b is the same, which reads

$$\boldsymbol{p} = 3\varepsilon_{eff}V\frac{\varepsilon_b - \varepsilon_{eff}}{\varepsilon_b + 2\varepsilon_{eff}}\mathbf{E}. \tag{2.33}$$

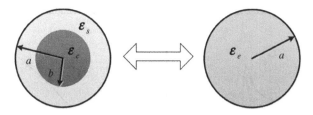

Figure 2.8 The coated sphere and its equivalent counterpart.

Then, for the actual situation where both a and b are present, the induced dipole moment inside the circle can be expressed as the weighted sum of the two factors:

$$p = 3\varepsilon_{eff}V\frac{\varepsilon_a - \varepsilon_{eff}}{\varepsilon_a + 2\varepsilon_{eff}}E*f_a + 3\varepsilon_{eff}V\frac{\varepsilon_b - \varepsilon_{eff}}{\varepsilon_b + 2\varepsilon_{eff}}E*f_b. \tag{2.34}$$

However, as we all know, the internal and external parts are exactly the same, which suggests there should be no additional dipole moment inside the circle (as a result, there will be no field deviations outside the circle), that is,

$$\frac{\varepsilon_a - \varepsilon_{eff}}{\varepsilon_a + 2\varepsilon_{eff}}*f_a + \frac{\varepsilon_b - \varepsilon_{eff}}{\varepsilon_b + 2\varepsilon_{eff}}*f_b = 0. \tag{2.35}$$

This formula is Bruggeman's equation in the symmetrical form. It is clear that the two media are equivalent in Eq. (2.35). The interchange of the two terms will lead to the same effective value, which explains the name of the symmetrical formula.

2.5.5 Internal Homogenization Techniques

Sometimes, the inclusions in inhomogeneous media have their own internal structures (e.g., they may have several layers, like coated spheres). Internal homogenization, as proposed by Engheta et al., means that one assigns an effective permittivity for a single inclusion instead of the whole medium, which they call the external homogenization [22].

Fig. 2.8 shows a coated sphere and its equivalent counterpart, a sphere with the same radius. Both spheres are in deep subwavelength scales and hence can be treated in the quasi-static way. Suppose that a uniform electric field ($\vec{E}_0$) is exerted on the spheres, which is in the z direction. Then, using the classical method of separation of variables under the spherical coordinate system, together with boundary conditions at the interface, one can easily get the field distributions outside the spheres.

For the coated sphere, the electric potentials in the core, shell, and ambient environment can be expressed as follows:

$$\phi_c = ArP_1(\cos\theta), \tag{2.36a}$$

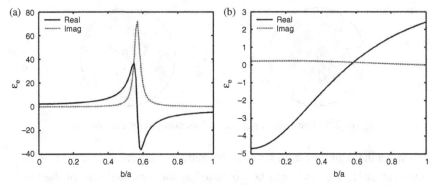

Figure 2.9 The effective permittivity by using two materials under two configurations. (a) $\varepsilon_c = -4.70 + i0.22$, $\varepsilon_s = 2.42$, (b) $\varepsilon_s = -4.70 + i0.22$, $\varepsilon_c = 2.42$.

$$\phi_s = [Br + Cr^{-2}]P_1(\cos\theta), \tag{2.36b}$$

$$\phi_a = [-E_0r + Dr^{-2}]P_1(\cos\theta), \tag{2.36c}$$

which can be solved using separation of variables. For its counterpart, the electric potentials are written as

$$\phi_c = A'rP_1(\cos\theta), \tag{2.37a}$$

$$\phi_a = [-E_0r + D'r^{-2}]P_1(\cos\theta). \tag{2.37b}$$

If the two particles are said to be equivalent, then it is quite natural that they should give the same potential distributions under the same excitations. In our case, D and D' should be equal. The equality of these coefficients gives the following formula:

$$\varepsilon_e = \varepsilon_s\frac{a^3(\varepsilon_c + 2\varepsilon_s) + 2b^3(\varepsilon_c - \varepsilon_s)}{a^3(\varepsilon_c + 2\varepsilon_s) - b^3(\varepsilon_c - \varepsilon_s)}. \tag{2.38}$$

A comparison of this equation with Eq. (2.26) shows that it is again the M–G formula. However, it is obtained using a different method.

The approach also applies for other inclusions. For example, it is applicable for 2D coated cylinders. (Refer to Fig. 2.8 for the cross-sectional view of the structure.) In this scenario, a similar equation holds, which reads

$$\varepsilon_e = \varepsilon_s\frac{a^2(\varepsilon_c + \varepsilon_s) + b^2(\varepsilon_c - \varepsilon_s)}{a^2(\varepsilon_c + \varepsilon_s) - b^2(\varepsilon_c - \varepsilon_s)}. \tag{2.39}$$

It should be pointed out that the internal homogenization scheme has been used by the same researchers in their proposal for digital metamaterials, where a coated cylinder is used as a "digital bit" for the formation of a "digital byte" and metamaterials [23]. Fig. 2.9 demonstrates the effective permittivity using the scheme with two different configurations. It is clear that this simple combination can give rise to a wide range of effective permittivities.

2.6 Summary

So far we have shown various classical models for electromagnetically characterizing natural materials. The typical rules are also given when two or more materials are mixed together. Please note that these theories apply for material permeability as well as permittivity. With knowledge of the theories just presented, the analysis and realization of metamaterials will be much easier.

3 Retrieval Methods of Effective Medium Parameters

In this section, we introduce and discuss the retrieval methods of effective medium parameters for metamaterials. First, we present the popular retrieval method of the effective permittivity and permeability from scattering parameters. Then we discuss a special case when the real part of impedance $Re(z)$ and the imaginary part of refractive index $Im(n)$ are close to zero, in which the regular method may fail and we need to handle the signs of n and z. In most designs of metamaterials for complicated functionalities, bianisotropy and inhomogeneity are usually used and are very important. Hence, in Sections 3.3 and 3.4, we investigate the retrieval methods of bianisotropic and inhomogeneous metamaterials, respectively.

3.1 Regular Retrieval Method

As we discussed in previous sections, whether the subwavelength-scale elements are periodically or nonperiodically distributed, metamaterials can always be equivalent to continuous media [1]. Many experiments on metamaterials have shown that the EM field distributions in the subwavelength structures are indeed consistent with materials having the equivalent electric permittivity and magnetic permeability [4, 7, 24–26]. In this section, we introduce the retrieval method for the equivalent medium parameters proposed by Smith et al. in 2002 [27]. By using this theory, we extract the refractive index and impedance from the scattering S-parameters (transmission coefficients and reflection coefficients) of the structure, and hence the equivalent permittivity and permeability can be obtained. There are some other extraction methods for the medium parameters, such as quasi-static field analysis [28], the equivalent-circuit method [29–33], and so on. These theories are important for the design of artificial elements and the experimental verification of metamaterials.

It is well known that the impedance (z) and refractive index (n) are always used to describe the EM properties of homogeneous materials. In fact, when the scattering properties are described by these two quantities, another group of variables are more convenient to obtain: the electric permittivity $\varepsilon = n/z$ and magnetic permeability $\mu = nz$. Both n and z, or ε and μ, are complex functions related to the frequencies. Generally, for passive materials, Re(z) and Im(n) are greater than zero.

When the EM wave is vertically incident to an infinitely large and homogeneous dielectric slab with thickness d in free space, the transmission coefficient can be expressed by n and z as

$$S_{12} = \frac{1}{\cos(nk_0 d) - \frac{i}{2}(z + \frac{1}{z})\sin(nk_0 d)}, \tag{3.1}$$

where $k_0 = \omega/c$ denotes the wave number of the incident wave. We consider the case in which the incident wave propagates along the x direction and the origin of the coordinate is the incident end of the material. Then, based on z and n, the reflection coefficient is expressed as

$$S_{11} = -\frac{i}{2}(z - \frac{1}{z})\sin(nk_0 d). \tag{3.2}$$

According to Eqs. (3.1) and (3.1), z and n can be expressed by using S_{11} and S_{12} as

$$\cos(nk_0 d) = \frac{1}{2S_{21}}(1 + S_{21}^2 - S_{11}^2) \tag{3.3}$$

and

$$z = \pm\sqrt{\frac{(1 + S_{11})^2 - S_{21}^2}{(1 - S_{11})^2 - S_{21}^2}}, \tag{3.4}$$

which provide complete parameter expressions of a slab composed of homogeneous material. Note that Eqs. (3.1) and (3.2) satisfy the energy-conservation law, that is to say, $|S_{11}|^2 + |S_{12}|^2 = 1$, if the metamaterial is lossless.

Although the expressions of z and n are not complicated, they are multivalued functions of complex variables. Hence we need additional conditions to determine the correct values of ε and μ. For example, if the material is passive, the sign in Eq. (3.4) can be determined by the condition

$$\text{Re}(z) > 0. \tag{3.5}$$

Similarly, the value of Im(n) can be determined by the condition

$$\text{Im}(n) > 0, \tag{3.6}$$

which leads to an unambiguous result for Im(n):

$$\text{Im}(n) = \pm\text{Im}(\frac{\cos^{-1}([1 - (S_{11}^2 - S_{21}^2)]/(2S_{21}))}{k_0 d}). \tag{3.7}$$

Re(n) can be calculated by

$$\text{Re}(n) = \pm\text{Re}\left(\frac{\cos^{-1}([1 - (S_{11}^2 - S_{21}^2)]/(2S_{21}))}{k_0 d}\right) + \frac{2\pi m}{k_0 d}, \tag{3.8}$$

which is complicated by the branches of the arc cosine function (m is an integer). It should be noticed that when d is very large, the values of these branches are close to each other, which makes it more difficult to choose the correct value in the dispersive materials. For this reason, the thickness is often chosen as small as possible, such as one-element thickness, and hence metamaterials can be analyzed the same way as continuous materials [4, 7, 24–26].

3.2 Removal of Singularity

In the general case, the material parameters can be retrieved from the S-parameters. However, when the S-parameters obtained from experimental measurements or numerical simulations are noisy, the retrieval method may fail, especially at those frequencies where z and n are sensitive to small variations of S_{11} and S_{21}. This problem has been addressed and discussed in detail in Ref. [34].

As discussed in Section 3.1, one can usually determine z and n from Eqs. (3.1) and (3.2) with the requirement of Eqs. (3.5) and (3.6), where z and n are determined independently. However, this method may fail in practice when Re(z) and Im(n) are close to zero, in which case a slight perturbation of S_{11} and S_{21} may change the signs of Re(z) and Im(n), making it unreliable to apply the requirements of Eqs. (3.5) and (3.6). Actually, z and n are related and their relationship can be used to determine the signs in Eqs. (3.4), (3.7), and (3.8). In order to determine the correct sign of z, two cases were discussed mathematically in Ref. [34]. The first is for $|\text{Re}(z)| \geq \delta_z$, where δ_z is a positive number. In such a case, Eq. (3.5) can be applied. In the second case of $|\text{Re}(z)| < \delta_z$, the sign of z is determined to ensure that the corresponding refractive index n has a non-negative imaginary part, or equivalently $|e^{ink_0 d}| \leq 1$, where n can be derived from Eqs. (3.1) and (3.2):

$$e^{ink_0 d} = \frac{S_{21}}{1 - S_{11}\frac{z-1}{z+1}}. \tag{3.9}$$

Note that once we obtain the value of z, the value of $e^{ink_0 d}$ is achieved from Eq. (3.9), and hence we avoid the sign ambiguity of Im(n) in Eq. (3.7). Fig. 3.1(b) shows the retrieved impedance of split-ring-resonator (SRR)-wire structure, as illustrated in Fig. 3.1(a), using the method presented in Ref. [34] and using only the condition of Eq. (3.5). We remark that the discontinuities obtained by only applying the criterion Re(z) ≥ 0 have been removed.

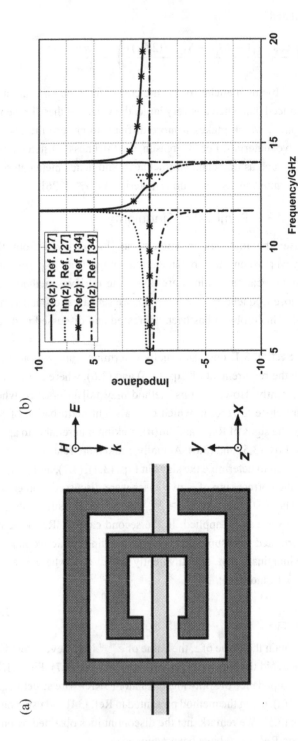

Figure 3.1 (a) Illustration of a metamaterial element. The side-lengths of SRR-wire structure along the x and z directions are $a_x = 4$ mm, $a_z = 3$ mm. The element thickness d_0 in the direction of wave incidence is 4 mm. (b) Comparison of the retrieved impedance z (real and imaginary parts) for the metamaterial element by the method presented in Refs. [27] and [34].

Next we discuss $\mathrm{Re}(n)$, which remains ambiguous due to the branches of the arccosine function in Eq. (3.8). In order to address this problem, Chen et al. proposed a mathematical method to determine the proper branch by using the continuity of the parameters [34], with special attention to possible discontinuities due to resonances. That is an iterative method: assuming that the value of refractive index $n(f_0)$ at a frequency f_0 has been obtained, $n(f_1)$ at the next frequency sample f_1 can be obtained by expanding the function $e^{in(f_1)k_0(f_1)d}$ in a Taylor series,

$$e^{in(f_1)k_0(f_1)d} \approx e^{in(f_0)k_0(f_0)d}(1 + \Delta + \frac{1}{2}\Delta^2), \tag{3.10}$$

where $\Delta = in(f_1)k_0(f_1)d - in(f_0)k_0(f_0)d$ and $k_0(f_0)$ denotes the wave number in free space at the frequency f_0.

In Eq. (3.10), the branch index of $\mathrm{Re}[n(f_1)]$ is the only unknown. The left side of Eq. (3.10) is obtained from Eq. (3.9), and hence Eq. (3.10) is a binomial function of the unknown $n(f_1)$. Between the two roots, one is an approximation of the true solution. Because we have obtained $\mathrm{Im}[n(f_1)]$, we can choose the correct root between the two by comparing the imaginary parts with $\mathrm{Im}[n(f_1)]$. The root with imaginary part closer to $\mathrm{Im}[n(f_1)]$ is the correct one, and we denote it as n_0. Since n_0 is a good approximation to $n(f_1)$, the integer m can be chosen in Eq. (3.8) so that $\mathrm{Re}(n(f_1))$ is as close to $\mathrm{Re}(n_0)$ as possible.

The branch of $\mathrm{Re}(n)$ at the initial frequency is determined as follows:

$$\mathrm{Im}(\mu) = \mathrm{Re}(n)\mathrm{Im}(z) + \mathrm{Im}(n)\mathrm{Re}(z), \tag{3.11}$$

$$\mathrm{Im}(\varepsilon) = \frac{1}{|z|^2}(-\mathrm{Re}(n)\mathrm{Im}(z) + \mathrm{Im}(n)\mathrm{Re}(z)). \tag{3.12}$$

The requirements $\mathrm{Im}(\mu) \geq 0$ and $\mathrm{Im}(\varepsilon) \geq 0$ lead to

$$|\mathrm{Re}(n)\mathrm{Im}(z)| \leq \mathrm{Im}(n)\mathrm{Re}(z). \tag{3.13}$$

In particular, when $\mathrm{Im}(n)\mathrm{Re}(z)$ is close to zero but $\mathrm{Im}(z)$ is not, $\mathrm{Re}(n)$ should be close to zero. At the initial frequency, we solve for the branch integer m satisfying Eq. (3.13). If there is only one solution, it is the correct branch. In case of multiple solutions, for each candidate branch index m, the value of $\mathrm{Re}(n)$ can be determined at all subsequent frequencies using the above-mentioned method. Because the requirement of Eq. (3.13) applies to $\mathrm{Re}(n)$ at all frequencies, we can use it to verify $\mathrm{Re}(n)$ at all frequencies produced by the candidate initial branch. Note that in the special case when $\mathrm{Im}(n)\mathrm{Re}(z)$ is close to zero while $\mathrm{Im}(z)$ is not, the checking process can easily be carried out. Therefore, the initial branch that satisfies Eq. (3.13) at both the initial frequency and subsequent frequencies is the correct one.

3.3 Constitutive Parameters of Bianisotropic Metamaterial

In this section, we consider the bianisotropy of the metamaterials [35]. We also take a commonly used edge-coupled SRR as an example, which is illustrated in Fig. 3.2. The SRR structure is composed of two concentric metallic rings that are both interrupted by a small gap. When a plane wave is incident in the x direction with the electric field in the z direction, SRR will present a bianisotropic property. The reason is that the electric field in the z direction can induce a magnetic dipole in the y direction due to the asymmetry of the inner and outer rings, while the magnetic field in the y direction can induce an electric dipole in the z direction. By assuming that the medium is reciprocal and the harmonic time dependence is $e^{-i\omega t}$, the constitutive relationships can be expressed as

$$\boldsymbol{D} = \bar{\bar{\varepsilon}} \cdot \boldsymbol{E} + \bar{\bar{\xi}} \cdot \boldsymbol{H}, \tag{3.14}$$

$$\boldsymbol{B} = \bar{\bar{\mu}} \cdot \boldsymbol{H} + \bar{\bar{\zeta}} \cdot \boldsymbol{E}, \tag{3.15}$$

where $\bar{\bar{\varepsilon}} = \varepsilon_0(\varepsilon_x, \varepsilon_y, \varepsilon_z)$, $\bar{\bar{\mu}} = \mu_0(\mu_x, \mu_y, \mu_z)$, and

$$\bar{\bar{\xi}} = \frac{1}{c} \begin{pmatrix} 0 & 0 & 0 \\ 0 & 0 & 0 \\ 0 & -i\xi_0 & 0 \end{pmatrix}, \tag{3.16}$$

$$\bar{\bar{\zeta}} = \frac{1}{c} \begin{pmatrix} 0 & 0 & 0 \\ 0 & 0 & i\xi_0 \\ 0 & 0 & 0 \end{pmatrix}. \tag{3.17}$$

Here, ε_0 and μ_0 are the permittivity and permeability in the free space, and c is the speed of light in free space. The seven unknowns ε_x, ε_y, ε_z, μ_x, μ_y, μ_z, and ξ_0 are quantities without dimensions. When a plane wave polarized in the z direction is incident along the x direction, the three parameters ε_z, μ_y, and ξ_0 will be active, while the other four parameters ε_x, ε_y, μ_x, and μ_z will not be involved in the bianisotropic process.

Note that the characteristic impedances have different values for a bianisotropic material when the waves propagate in two opposite directions of the x-axis. For an EM wave propagating in the x direction, the impedances can be expressed as

$$z^+ = \frac{\mu_y}{n + i\xi_0}, \quad z^- = \frac{\mu_y}{n - i\xi_0}, \tag{3.18}$$

respectively. Here n is the effective refractive index that has the same value for the EM wave propagating in the opposite direction of the x-axis,

$$n = \pm\sqrt{\varepsilon_z\mu_y - \xi_0^2}. \tag{3.19}$$

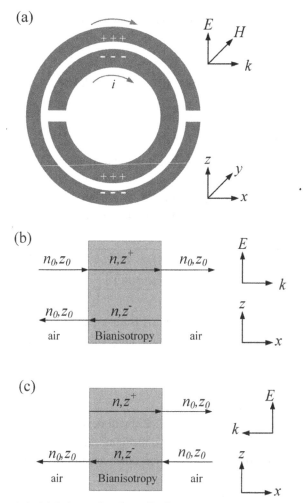

Figure 3.2 (a) Schematic of a split-ring resonator used to construct metamaterials. When a plane wave polarized along the *z*-axis is incident in the *x* direction, the metamaterial will show bianisotropy. (b–c) Schematics of a homogeneous bianisotropic slab placed in open space for the calculation of S-parameters. (b) and (c) are for plane waves incident in the +*x* and −*x* directions, respectively.

As discussed in Section 3.1, a periodic metamaterial can be approximated as a homogeneous medium under the condition of a long wavelength. Hence, a simplified model of a bianisotropic material slab can be used to calculate the S-parameters. Fig. 3.2 illustrates a homogeneous bianisotropic material slab that is placed in an open space. Two situations are considered (i.e., incidences along the +*x* and −*x* directions). By applying the continuous boundary conditions, the expressions of S-parameters can be obtained by using the transfer matrix method [35]. When the incidence is along the +*x* direction as shown in Fig. 3.2(b), the corresponding reflection coefficient (S_{11}) and transmission coefficient (S_{21}) are calculated as

$$S_{11} = \frac{2i\sin(nk_0 d)[n^2 + (\xi_0 + i\mu_y)^2]}{[(\mu_y + n)^2 + \xi_0^2]e^{-ink_0 d} - [(\mu_y - n)^2 + \xi_0^2]e^{ink_0 d}}, \tag{3.20}$$

$$S_{12} = \frac{4\mu_y n}{[(\mu_y + n)^2 + \xi_0^2]e^{-ink_0 d} - [(\mu_y - n)^2 + \xi_0^2]e^{ink_0 d}}, \tag{3.21}$$

where d is the thickness of the bianisotropic metamaterial slab and k_0 is the wave number in the free space. For the case when the incidence is along the $-x$ direction, as shown in Fig. 3.2(c), the corresponding reflection coefficient (S_{22}) and transmission coefficient (S_{12}) can be obtained as

$$S_{21} = \frac{4\mu_y n}{[(\mu_y + n)^2 + \xi_0^2]e^{-ink_0 d} - [(\mu_y - n)^2 + \xi_0^2]e^{ink_0 d}}, \tag{3.22}$$

$$S_{22} = \frac{2i\sin(nk_0 d)[n^2 + (\xi_0 - i\mu_y)^2]}{[(\mu_y + n)^2 + \xi_0^2]e^{-ink_0 d} - [(\mu_y - n)^2 + \xi_0^2]e^{ink_0 d}}. \tag{3.23}$$

It is observed that S_{21} equals S_{12}, but S_{11} is not equal to S_{22}. Therefore, three independent equations can be used to solve three unknowns (n, μ_y, and ξ_0). The refractive index n is expressed as

$$\cos(nk_0 d) = \frac{1 - S_{11}S_{22} + S_{21}^2}{2S_{21}}. \tag{3.24}$$

Obviously, when S_{11} equals S_{22}, Eq. (3.22) will degenerate into Eq. (3.3). As discussed in Section 3.1, for a passive medium, the solved n must obey the condition of Eq. (3.6). After n is obtained, other constitutive parameters can be achieved:

$$\xi_0 = \left(\frac{n}{-2\sin(nk_0 d)}\right)\left(\frac{S_{11} - S_{22}}{S_{21}}\right), \tag{3.25}$$

$$\mu_y = \left(\frac{in}{\sin(nk_0 d)}\right)\left(\frac{2 + S_{11} + S_{22}}{2S_{21}} - \cos(nk_0 d)\right), \tag{3.26}$$

$$\varepsilon_z = \frac{n^2 + \xi_0^2}{\mu_y}. \tag{3.27}$$

Consequently, the impedances z^+ and z^- can be derived from Eq. (3.18). Again, for a passive medium, the following conditions should be satisfied:

$$\text{Re}(z^+) \geq 0, \ \text{Re}(z^-) \geq 0. \tag{3.28}$$

Hence, all of the bianisotropic constitutive parameters have been retrieved.

We take two single elements of the metamaterials; for example, as shown in Fig. 3.3. The incident wave is a plane wave with its wave vector k in the x direction and the E field polarized in the z direction. Fig. 3.3(a) is a metamaterial that is composed of a pure SRR with its gap opened in the z direction. This metamaterial is denoted as SRR-I and does not show bianisotropy for such incidence. Fig. 3.3(b) is a metamaterial that is composed of a pure SRR with its gap opened in the x direction, which is denoted as SRR-II, showing the bianisotropic properties.

Based on the S-parameters, we retrieve the constitutive parameters for the elements SRR-I and SRR-II, and the effective parameters ε_z, μ_y, and ξ_0 are shown in Figs. 3.3(d–f). For ε_z, the most significant difference is that Re(ε_z) of SRR-I demonstrates antiresonant behavior, while Re(ε_z) of SRR-II shows normal resonant behavior. For Re(μ_y), one can observe that the curve of SRR-II has a redshift compared to that of SRR-I. Moreover, the band of negative Re(μ_y) of SRR-II is shallower than that of SRR-I. Fig. 3.3(f) illustrates the magnetoelectric coupling parameter ξ_0. Evidently, ξ_0 of SRR-I is zero, while ξ_0 of SRR-II shows antiresonant behavior. However, this antiresonant behavior can be changed into a resonant behavior if we simply reverse the orientation of the SRR structure in the x direction. All of the retrieval results demonstrate the intrinsic differences between a normal element SRR-I and a bianisotropic element SRR-II.

3.4 Effective Parameters of Inhomogeneous Metamaterials

In the process of S-parameter retrievals presented in Sections 3.1 through 3.3, we always assume that all elements in a metamaterial slab exhibit the same response to the applied field, but this is only an approximation due to the existing coupling. The dissimilarity of coupling effects between edge elements and inner elements in a metamaterial slab has been considered in Ref. [36], and advanced parameter retrievals are presented based on an inhomogeneous medium model. The results demonstrate that the effective medium parameters for edge and inner elements are different and verify the inhomogeneous property of the metamaterial slab. The retrieval methods have been presented to obtain the effective medium parameters of edge and inner elements by considering two- and three-element metamaterial slabs (in the propagation direction), respectively, and an inhomogeneous three-layer model is also discussed for arbitrary inhomogeneous metamaterial slabs.

As is well known, the coupling between adjacent metamaterial elements always occurred. The elements of an infinitely large metamaterial slab can be categorized into two types, the edge element and the inner element, as shown in Fig. 3.4(a). The edge element couples with another element at the right or left side, while the inner element is affected by elements at both sides. Hence, a metamaterial slab is characterized into two-edge elements and a number of inner

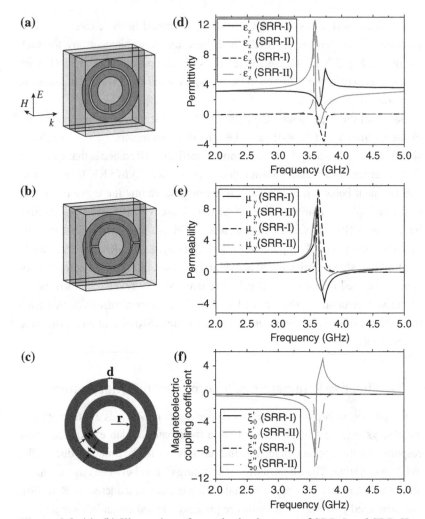

Figure 3.3 (a)–(b) Illustration of two single elements of SRR-I and SRR-II, respectively. (c) The detailed geometry of the SRR. The geometric parameters are $d = t = 0.2$ mm, $w = 0.9$ mm, and $r = 1.6$ mm. (d) Retrieved permittivity, (e) permeability, and (f) magnetoelectric coupling coefficient for the elements of SRR-I and SRR-II, respectively [35].

elements. In this section, we still take SRR structures for example, and present the process to achieve the medium parameters of inhomogeneous elements.

When an EM wave is normally incident on a metamaterial slab, the slab can be effectively considered as cascaded metamaterial layers, each of which has one element in the propagation direction, as shown in Fig. 3.4(a). The cascaded layers can be described by the multiplication of transmission matrices of all layers,

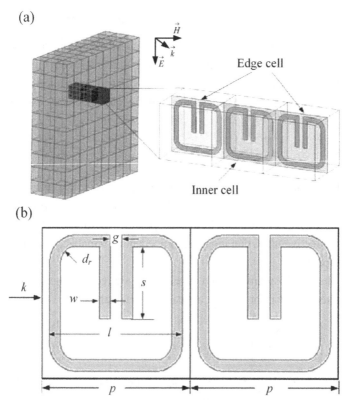

Figure 3.4 (a) Illustration of a metamaterial slab, where the edge and inner elements are shown in the right panel. (b) A two-element metamaterial slab composed of SRRs.

$$(T) = (T_1)(T_2) \cdots (T_n), \tag{3.29}$$

where T is the transmission matrix of the whole slab,

$$(T) = \begin{bmatrix} T_{11} & T_{12} \\ T_{21} & T_{22} \end{bmatrix}, \tag{3.30}$$

which can be translated to S-parameters by

$$(S) = \frac{1}{T_{11}} \begin{bmatrix} T_{21} & |T| \\ 1 & -T_{12} \end{bmatrix}, \quad (T) = \frac{1}{S_{21}} \begin{bmatrix} 1 & -S_{22} \\ S_{11} & -|S| \end{bmatrix}. \tag{3.31}$$

Next we discuss the retrieval of effective medium parameters for the edge and inner elements, respectively, by considering two- and three-element metamaterial slabs.

3.4.1 Parameter Retrieval of Edge Element

Consider a two-element metamaterial slab composed of SRR in which both elements are the edge ones, as shown in Fig. 3.4(b). When a wave is incident along the z direction, the magnetic field is perpendicular to the plane of the rings. Suppose that the normalized S-parameters of the left element are expressed as

$$(S') = \begin{bmatrix} S'_{11} & S'_{12} \\ S'_{21} & S'_{22} \end{bmatrix}, \tag{3.32}$$

in which the superscript "*l*" corresponds to the left ring, $S'_{11} = \frac{z-1}{z+1}$, $S'_{12} = \frac{2}{z+1}e^{ik_0np}$, $S'_{21} = \frac{2z}{z+1}e^{ik_0np}$, $S'_{22} = -\frac{z-1}{z+1}e^{i2k_0np}$, and $k_0 = \omega/c$ is the wave number of the incident wave in free space, p is the length of an element, and n and z are effective refraction index and impedance normalized to those in free space, respectively.

Since the slab is symmetric and the elements are reciprocal, the normalized S-parameters of the right element are easily written as

$$(S^r) = \begin{bmatrix} S'_{22} & S'_{21} \\ S'_{12} & S'_{11} \end{bmatrix}, \tag{3.33}$$

in which the superscript "*r*" corresponds to the right ring. Substituting Eqs. (3.33) and (3.32) into Eq. (3.31), the transmission matrices of two elements are obtained,

$$(T_l) = \frac{1}{S'_{21}} \begin{bmatrix} 1 & -S'_{22} \\ S'_{11} & -|S'| \end{bmatrix}, \quad (T_r) = \frac{1}{S'_{12}} \begin{bmatrix} 1 & -S'_{11} \\ S'_{22} & -|S^r| \end{bmatrix}, \tag{3.34}$$

which relates to the S-parameter of the whole slab by the following equation:

$$(T_l)(T_r) = \frac{1}{S_{21}} \begin{bmatrix} 1 & -S_{22} \\ S_{11} & -|S| \end{bmatrix}. \tag{3.35}$$

By solving Eqs. (3.32) through (3.35), the S-parameters (S_{11} and S_{21}) of the two elements can be obtained from simulated or measured data (S_{11} and S_{21}). Applying the method in Section 3.1, the material parameters of the edge elements can be obtained.

3.4.2 Parameter Retrieval of Inner Element

Next, we consider a three-element metamaterial slab, as shown in Fig. 3.4(a), in which the two elements on both sides are edge elements and the middle one is an inner element. The incident wave propagates along the z direction with the magnetic field perpendicular to the rings. The transmission matrix of the slab, T, is expressed as

$$(T) = (T_l)(T_m)(T_r), \tag{3.36}$$

where T_m is the transmission matrix of the middle element. Since T is available from the simulated or experimental S-parameters, then T_m can be achieved using the results of edge elements by

$$(T_m) = (T_l)^{-1}(T)(T_r)^{-1}, \tag{3.37}$$

in which T_l and T_r are described in Eq. (3.34). Hence, the S-parameters of the inner element can be easily computed from Eqs. (3.31) and (3.37). Based on the retrieval method presented in Section 3.1, the effective refraction index and normalized impedance for the inner element are easily obtained.

Next, we discuss the effective parameters of the SRR structures shown in Fig. 3.4 based on the method shown in Section 3.1 and the method shown in Section 3.4, respectively. The geometric parameters are chosen as $p = 3.3333$ mm, $l = 3$ mm, $w = g = 0.25$ mm, $s = 1.6$ mm, and $d_r = 0.35$ mm. The effective medium parameters of a single SRR without considering any coupling are shown in Fig. 3.5(a). Fig. 3.5(b) gives the retrieved medium parameters for the edge element of SRR, and the corresponding results for the inner SRR element are illustrated in Fig. 3.5(c). Comparing Figs. 3.5(b) and 3.5(c) with Fig. 3.5(a), we clearly observe the differences of medium parameters between the edge and inner elements, which verify the effect of coupling between adjacent elements to EM parameters. Such differences are due to different couplings of neighboring elements. With the effective medium parameters of edge and inner elements distinguished, most coupling effects have been taken into consideration, and the element responses in metamaterial slabs can be identified more accurately by using the method in Ref. [36]. Observe there are two peaks in Fig. 3.5(c). The possible reason is that the discontinuity at the edge excites unwanted surface waves and distorts the Bloch–Fouquet waves, and hence the second resonant peak arises.

For an arbitrary N-element metamaterial slab ($N = 3, 4, 5, ...$), a three-layer inhomogeneous medium model can be set up based on the edge and inner elements. In the three-layer model, the first layer is for the left edge element with the thickness p, the second layer is for $N - 2$ inner elements with the thickness $(N - 2)p$, and the third layer is for the right edge element with the thickness p. The effective medium parameters in the first and third layers can be obtained based on the information presented in Section 3. After substituting the retrieved material parameters of the inner and edge elements into the

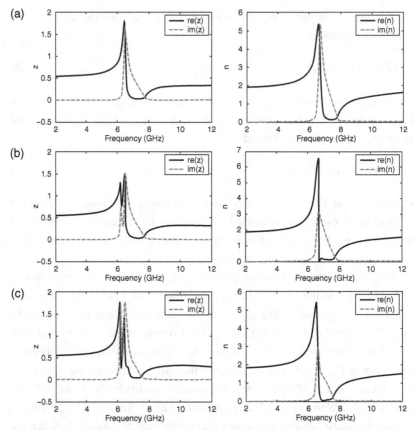

Figure 3.5 The relative impedance and refraction index parameters of (a) a single SRR element, (b) an edge SRR element, and (c) an inner SRR element. The solid lines are real parts and the dashed lines are imaginary parts.

corresponding layers, the total transmission coefficient T and reflection coefficient R can be analytically calculated by using the forward-propagating matrix [37]. The three-layer inhomogeneous medium model provides a simple description for the metamaterial slab [36].

3.5 Summary

In this section, we have introduced and discussed the retrieval methods of effective medium parameters for metamaterial elements. We first discussed the retrieval method of single elements and then particularly considered the case when $\mathrm{Re}(z)$ and $\mathrm{Im}(n)$ are close to zero, in which we need to determine the sign choice of n and z. In most designs of metamaterials, bianisotropy and inhomogeneity are very useful. Hence, we also presented the retrieval methods of bianisotropic and inhomogeneous metamaterials, respectively.

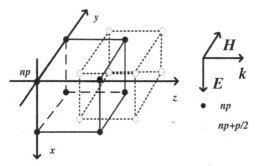

Figure 4.1 Grids used in the analysis. Two sets of interleaved grids are involved, one represented by the red (black) color and the other by the dark blue (grey) color. The periodicity of both grids is p.

4 Homogenization of Metamaterials by Field Averaging

The retrieval process based on S-parameters provides a very effective way for determining equivalent permittivity and permeability for metamaterials, which treats a metamaterial slab as a natural material slab and builds the equivalence using S-parameters [38]. This process can be considered a "black box" method, meaning the materials in the box may be totally different (material or metamaterial) but have the same EM responses outside the box. Though very effective and convenient, the method does not provide any physical interpretation of the metamaterial and usually involves a time-consuming process for the S-parameters' calculation. In this regard, the field averaging method is introduced to eliminate these disadvantages.

4.1 Definition of the Averaged Fields

As discussed in the previous sections, natural materials constitute billions of billions of atoms, which are fixed at grid points with different lattice structures, and the calculation of their permittivity/permeability will definitely involve a field averaging process so as to eliminate the large fluctuation of local fields. Likewise, metamaterials can also be considered as an artificial crystal, with "artifical atoms" (electric/magnetic resonator) situated at each grid point. Then, it seems quite natural that we can get the permittivity and permeability of metamaterials using the same methodology (i.e., by field averaging).

To begin with, let's first look at the grid structures of a metamaterial, however, we do not care which particular "atoms" are used in this structure. Hence, this process applies to all metamaterial structures. Fig. 4.1 shows a typical unit cell structure, whose period is p in three directions, indicated by the red-colored (black-colored for the black and white figure) cube. We apply Maxwell's equations on this particular cube using its integral form, which are

$$\oint \vec{E} \cdot d\vec{l} = -i\omega \int\int \vec{B} \cdot d\vec{S}, \tag{4.1}$$

$$\oint \vec{H} \cdot d\vec{l} = i\omega \int\int \vec{D} \cdot d\vec{S}. \tag{4.2}$$

Please note that Eqs. (4.1) and (4.2) are vectorial and involve three scalar ones, which are easily represented by the fields on three surfaces of the cube and its corresponding edges. For example, the equation on the *xoz* surface can be identified as

$$\int_0^P E_x(x,0,0)dx - \int_0^P E_x(x,0,p)dx + \int_0^P E_z(p,0,z)dz - \int_0^P E_z(0,0,z)dz$$
$$= i\omega \int_0^p \int_0^p B_y(x,0,z)dxdz \tag{4.3}$$

Using Eq. (4.3), it seems natural to introduce the averaged electric field as follows:

$$\bar{E}_x(\tfrac{p}{2},0,0) = \frac{1}{p}\int_0^P E_x(x,0,0)dx, \ \bar{E}_x(\tfrac{p}{2},0,p) = \frac{1}{p}\int_0^P E_x(x,0,p)dx, \tag{4.4}$$

$$\bar{E}_z(p,0,\tfrac{p}{2}) = \frac{1}{p}\int_0^P E_z(p,0,z)dz, \ \bar{E}_z(0,0,\tfrac{p}{2}) = \frac{1}{p}\int_0^P E_z(0,0,z)dz, \tag{4.5}$$

which is defined at the middle point of the edge. And the averaged magnetic field is

$$\bar{B}_y(\tfrac{p}{2},0,\tfrac{p}{2}) = \frac{1}{p^2}\int_0^p \int_0^p B_y(x,0,z)dxdz, \tag{4.6}$$

which is defined at the surface center.

Then, using the averaged quantities, Eq. (4.3) can be rewritten as

$$\bar{E}_x(\tfrac{p}{2},0,0) - \bar{E}_x(\tfrac{p}{2},0,p) + \bar{E}_z(p,0,\tfrac{p}{2}) - \bar{E}_z(0,0,\tfrac{p}{2}) = i\omega p \bar{B}_y(\tfrac{p}{2},0,\tfrac{p}{2}). \tag{4.7}$$

Please note that these averaged quantities are defined at different grid points, especially for the magnetic flux density, which involves positions of half the periodicity. This fact requires that when using an integral equation for Ampere's law (i.e., Eq. (4.2)), one has to shift the grid for $p/2$ in three directions so that the same quantity is defined at the same position. To show this, we also draw the shifted grid points in Fig. 4.1 with a dark-blue (grey) color. Actually, the grid configuration is the same as that used in the FDTD method, which was proposed by Yee et. al. [39]. The only difference between the two is the relative size of the cell with respect to the working wavelength. Using the shifted grids, similar equations can be obtained for Ampere's law:

$$\int_{p/2}^{3p/2} H_z(\tfrac{p}{2},\tfrac{p}{2},z)dz - \int_{p/2}^{3p/2} H_z(\tfrac{p}{2},\tfrac{3p}{2},z)dz + \int_{p/2}^{3p/2} H_y(\tfrac{p}{2},y,\tfrac{3p}{2})dy$$
$$- \int_{p/2}^{3p/2} H_y(\tfrac{p}{2},y,\tfrac{p}{2})dy = -i\omega \int_{p/2}^{3p/2}\int_{p/2}^{3p/2} D_x(\tfrac{p}{2},y,z)dydz. \tag{4.8}$$

And the averaged quantities are

$$\bar{H}_z(\tfrac{p}{2},\tfrac{p}{2},p) = \frac{1}{p}\int_{p/2}^{3p/2} H_z(\tfrac{p}{2},\tfrac{p}{2},z)dz, \bar{H}_z(\tfrac{p}{2},\tfrac{3p}{2},p)$$

$$= \frac{1}{p}\int_{p/2}^{3p/2} H_z(\tfrac{p}{2},\tfrac{3p}{2},z)dz, \tag{4.9}$$

$$\bar{H}_y(\tfrac{p}{2},p,\tfrac{3p}{2}) = \frac{1}{p}\int_{p/2}^{3p/2} H_y(\tfrac{p}{2},y,\tfrac{3p}{2})dy, \bar{H}_y(\tfrac{p}{2},p,\tfrac{p}{2})$$

$$= \frac{1}{p}\int_{p/2}^{3p/2} H_y(\tfrac{p}{2},y,\tfrac{p}{2})dy, \tag{4.10}$$

$$\bar{D}_x(\tfrac{p}{2},p,p) = \frac{1}{p^2}\int_{p/2}^{3p/2}\int_{p/2}^{3p/2} D_x(\tfrac{p}{2},y,z)dydz. \tag{4.11}$$

Using averaged quantities, Eq. (4.8) can be expressed as

$$\bar{H}_z(\tfrac{p}{2},\tfrac{p}{2},p) - \bar{H}_z(\tfrac{p}{2},\tfrac{3p}{2},p) + \bar{H}_y(\tfrac{p}{2},p,\tfrac{3p}{2}) - \bar{H}_y(\tfrac{p}{2},p,\tfrac{p}{2}) = -i\omega p \bar{D}_x(\tfrac{p}{2},p,p). \tag{4.12}$$

Similarly, four additional equations like Eqs. (4.7) and (4.12) can be obtained, which represent Maxwell's equations by the averaged quantities. For the sake of clarity, they are not included in this section. Please refer to Ref. [40, 41] for more details.

4.2 Calculation of the Material Parameters

Now that we have Maxwell's equations represented by the averaged quantities, let's consider a simple case where a plane wave is propagating inside the metamaterial structure. Please refer to Fig. 4.1 for the polarization of the plane wave. In this case, Eqs. (4.7) and (4.12) can be further simplified as

$$\bar{E}_x(\tfrac{p}{2},0,0) - \bar{E}_x(\tfrac{p}{2},0,p) = i\omega p \bar{B}_y(\tfrac{p}{2},0,\tfrac{p}{2}), \tag{4.13}$$

$$\bar{H}_y(\tfrac{p}{2},p,\tfrac{3p}{2}) - \bar{H}_y(\tfrac{p}{2},p,\tfrac{p}{2}) = -i\omega p \bar{D}_x(\tfrac{p}{2},p,p). \tag{4.14}$$

Now, we can introduce the averaged permittivity and permeability as follows:

$$\bar{D}_x(\tfrac{p}{2},p,p) = \bar{\varepsilon}_x \bar{E}_x(\tfrac{p}{2},p,p), \tag{4.15}$$

$$\bar{B}_y(\tfrac{p}{2},0,\tfrac{p}{2}) = \bar{\mu}_y \bar{H}_y(\tfrac{p}{2},0,\tfrac{p}{2}), \tag{4.16}$$

and we have

$$\bar{\varepsilon}_x = \bar{D}_x(\tfrac{p}{2},p,p)/\bar{E}_x(\tfrac{p}{2},p,p), \tag{4.17}$$

$$\bar{\mu}_y = \bar{B}_y(\tfrac{p}{2},0,\tfrac{p}{2})/\bar{H}_y(\tfrac{p}{2},0,\tfrac{p}{2}). \tag{4.18}$$

The permittivity and permeability in other directions can be defined in the same manner.

Please note that the electromagnetic waves are actually propagating in a periodic structure (only one period is shown in the figure). Bloch's theorem holds for this scenario [42], which means

$$\vec{E}(x,y,z) = \vec{E}_0(x,y,z)\exp(-i\beta z), \tag{4.19}$$

$$\vec{H}(x,y,z) = \vec{H}_0(x,y,z)\exp(-i\beta z), \tag{4.20}$$

where $\vec{E}_0$ and $\vec{H}_0$ are periodic functions of the position with periodicity p in three directions, and β is the phase constant along the z direction. Taking this into consideration, Eqs. (4.13) and (4.14) are changed into

$$E_0(\tfrac{p}{2},0,0) - E_0(\tfrac{p}{2},0,0)e^{-i\beta p} = i\omega p \bar{\mu}_y H_0(\tfrac{p}{2},0,0)e^{-i\frac{\beta p}{2}}, \tag{4.21}$$

$$H_0(\tfrac{p}{2},0,0)e^{-i\frac{3\beta p}{2}} - H_0(\tfrac{p}{2},0,0)e^{-i\frac{\beta p}{2}} = -i\omega p \bar{\varepsilon}_x E_0(\tfrac{p}{2},0,0)e^{-i\beta p}. \tag{4.22}$$

Eliminating the common terms in these two equations, we get the following equation:

$$\sin^2\left(\frac{\beta p}{2}\right) = \omega^2 p^2 \bar{\varepsilon}_x \bar{\mu}_y/4. \tag{4.23}$$

This is the dispersion equation inside the metamaterial. When p is very small, then Eq. (4.23) can be approximated as

$$\beta = \omega\sqrt{\bar{\varepsilon}_x \bar{\mu}_y}, \tag{4.24}$$

which looks the same as that inside a natural material. However, when p is not very small, the accurate expression must be utilized.

Pendry and Smith also consider another case, where nothing is inside the cell [40], which means a plane wave propagating in vacuum. In this case, it is not difficult to get the permittivity and permeability from Eqs. (4.17) and (4.18):

$$\bar{\varepsilon}_x = \varepsilon_0 \frac{\sin(\beta p/2)}{(\beta p/2)}, \bar{\mu}_y = \mu_0 \frac{\sin(\beta p/2)}{(\beta p/2)}. \tag{4.25}$$

The calculations in this section reveal that the material parameters found by the field averaging method exhibit spatial dispersion (i.e., the material parameters are functions of the propagation vector).

When the cell is not empty, for example, there are electric or magnetic resonators inside the cell; then, similar results are obtained, as derived by Pendry et al. [40],

$$\bar{\varepsilon}_x = \varepsilon_0 \bar{\varepsilon}'_x \frac{\sin(\beta p/2)}{(\beta p/2)}, \bar{\mu}_y = \mu_0 \bar{\mu}'_y \frac{\sin(\beta p/2)}{(\beta p/2)}, \tag{4.26}$$

where $\bar{\varepsilon}'_x$ and $\bar{\mu}'_y$ are the "real" effective material parameters for the metamaterial concerned, and $\bar{\varepsilon}_x$ and $\bar{\mu}_y$ are the field-averaged quantities, which contain spatial dispersions and should be eliminated in the post-processing routines.

4.3 Simulation Results

When using the field averaging techniques for the determination of material parameters, one usually employs commercial software with an eigenmode solver. However, other configurations are also possible. Plane waves propagating along different directions can be used as basis functions to extract anisotropic parameters too [43–45]. For the eigenmode solver, only one unit cell is needed in the simulation, with periodic boundary conditions set. The surfaces along one axis of the unit cell are related with phase lags, which vary from 0 to 180. For a certain phase lag, the eigenfields inside the cell is numerically determined, which are then used to calculate the effective permittivity and permeability by using Eqs. (4.17) and (4.18). Since field averaging method always contains spatial dispersions, one has to eliminate its influence ($\frac{\sin(\beta p/2)}{\beta p/2}$) from the result by using Eq. (4.26).

There is another way to get the field averaging parameters, which, in our opinion, resembles that of the retrieval process using S-parameters. The method relies on the availability of two quantities (i.e., wave impedance z and the refractive index n). As described earlier, when a phase lag is set and the corresponding local fields are at hand, one can immediately know the effective

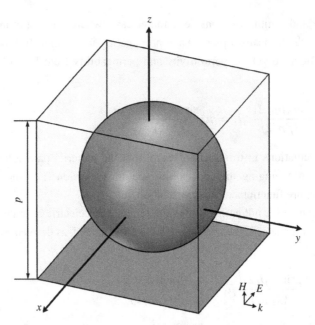

Figure 4.2 Unit cell structure for the artificial medium.

refractive index $n = \frac{\phi}{k_0 p}$, where $k_0 = \frac{\omega}{c}$; the wave impedance can also be obtained using $z = \frac{\bar{E}}{\bar{H}}$ with the averaged fields. Then, it is quite clear that

$$\bar{\varepsilon}' = n/z, \quad \bar{\mu}' = n \cdot z. \tag{4.27}$$

In the remainder of this section, we show an example by using the field averaging methods.

The artificial medium we consider consists of a cubic array of dielectric spheres, with a dielectric constant of 8. The lattice constant is 6 mm. Two configurations are considered, one for small spheres with radius $r = 0.1p$ and the other for big spheres with $r = 0.36p$. Fig. 4.2 shows one cell in our simulation. Eigenmode analysis is used in the numerical process. Then, the local field is averaged along the corresponding edges in the figure. After that, Eq. (4.27) is used to calculate the effective parameters.

Fig. 4.3 shows the calculated field-averaged results, where the horizontal axis represents the normalized frequency. The retrieved parameters based on reflection and transmission coefficients are also provided in the figure. It is quite clear that the two agree well with each other. Please note that equivalent parameters vary from 0.92 to 1.1, a very small interval, which explains the seemingly large descrepancy between the two.

Fig. 4.4 shows the results for $r = 0.36p$. Once again, excellent agreements are demonstrated in the figure. These results clearly confirm the correctness

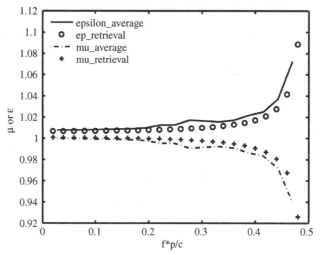

Figure 4.3 Comparison of field-averaged parameters with those from S-parameters, $r = 0.1p$. Note that c represents the speed of light in a vacuum and f is the frequency.

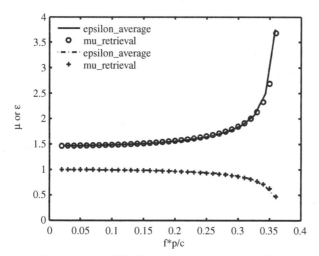

Figure 4.4 Comparison of field-averaged parameters with those from S-parameters, $r = 0.36p$.

of the field averaging method. It should be noted that an improvement for the method has also been proposed in the literature; please refer to Ref. [46, 47] for further readings and more simulations.

4.4 Summary

Generally speaking, the field averaging method involves firstly the numerical calculation of one unit cell in a metamaterial structure; then both the electric and magnetic fields are averaged along certain edges of the unit cell, and finally, the effective parameters are obtained using the averaged quantities. The field averaging method solves Maxwell's equations in a unit cell. It borrows the idea of "averaging" when analyzing natural materials to eliminate the fluctuation of local fields inside, and the results shown agree well with those obtained using retrieval methods. However, it should be noted that the method given in this section works in the quasi-static limit and does not apply directly for k-dependent or nonlocal metamaterials.

5 Effective Medium Theory of Metamaterials

Field averaging provides a natural and effective way for the calculation of parameters of metamaterials. However, as demonstrated in the previous section, the technique always bears the consequences of spatial dispersion, which hinders its wide application [41, 48]; while for the retrieval method based on S-parameters, the obtained permittivity and permeability seem to deviate from the standard Lorentz model, which we think they should strictly follow. In this section, we introduce an artificially defined wave impedance in the field averaging process, give the relation between the retrieved and averaged parameters, and thus bridge the gap between the two techniques. Our simulation results confirm the correctness of the theory.

5.1 Wave Impedance Artificially Defined in the Field Averaging Process

As shown in the previous section, for a metamaterial unit cell with periodicity p, the averaged parameters $\bar{\varepsilon}$ and $\bar{\mu}$ have the following connections with the phase delay across one unit structure:

$$\sin^2(\frac{\beta p}{2}) = \omega^2 p^2 \bar{\varepsilon}_x \bar{\mu}_y / 4,$$

which we now write as

$$\sin(\frac{\theta}{2}) = S_d \omega p \sqrt{\bar{\varepsilon}\bar{\mu}}/2, \tag{5.1}$$

where S_d denotes double positive or negative materials in the calculation. $S_d = 1$ for the former case and -1 for the latter case, and θ represents the phase lag across one cell.

Since there are two parameters in Eq. (5.1), another one must be found to uniquely determine the permittivity and permeability. And this is the artificially defined wave impedance in the unit cell. At first glance, this requirement seems easy because it is common knowledge that $\eta = \sqrt{\mu/\varepsilon}$. However, careful consideration will show that one has to accurately give the relative position in the cell for the definition, since the quantity varies along the propagation direction in the unit cell.

For example, if one calculates the impedance halfway along the propagation direction in one cell, then

$$\eta(np + p/2) = \bar{E}_x(np + p/2)/\bar{H}_y(np + p/2)$$

$$= \frac{2\bar{E}_x(np+p/2)}{\bar{H}_y(np)+\bar{H}_y(np+p)} = \sqrt{\frac{\bar{\mu}}{\bar{\varepsilon}}}\frac{1}{\cos(\theta/2)},$$

(5.2)

where Bloch's theorem is applied in the definition. However, if one defines the quantity at the cell surface, then

$$\eta(np) = \bar{E}_x(np)/\bar{H}_y(np) = \frac{\bar{E}_x(np-p/2)+\bar{E}_y(np+p/2)}{2\bar{H}_y(np)}$$

$$= \sqrt{\frac{\bar{\mu}}{\bar{\varepsilon}}}\cos(\theta/2).$$

(5.3)

Using the same expression as in Ref. [1, 38, 41, 49], we rewrite Eqs. (5.2) and (5.3) in one equation,

$$\eta = \sqrt{\frac{\bar{\mu}}{\bar{\varepsilon}}}(\cos\theta/2)^{S_b},$$

(5.4)

where $S_b = 1$ for electric resonators and -1 for magnetic resonators.

5.2 Connections between Retrieved and Averaged Parameters

Now, we have two sets of parameters in hand: one for the retrieved parameters using the S-parameters, represented by the subscript e (effective) in the expression, and the other for the averaged ones, denoted by the overhead bar in each expression, as listed in Section 5.1. As shown in the retrieval process, for the effective parameters we have

$$\frac{\omega}{c}p\sqrt{\varepsilon_e\mu_e} = \theta,$$

(5.5)

$$\eta = \sqrt{\frac{\mu_e}{\varepsilon_e}},$$

(5.6)

where c is the light speed in a vacuum, $c = \frac{1}{\sqrt{\varepsilon_0\mu_0}}$, and θ is the phase lag in one cell. If we combine Eqs. (5.1), (5.4), (5.5), and (5.6), then the effective parameters can be determined, expressed by the averaged quantities as

$$\varepsilon_e = \bar{\varepsilon}\frac{(\theta/2)}{\sin(\theta/2)}[\cos(\theta/2)]^{-S_b},$$

(5.7)

$$\mu_e = \bar{\mu}\frac{(\theta/2)}{\sin(\theta/2)}[\cos(\theta/2)]^{S_b}.$$

(5.8)

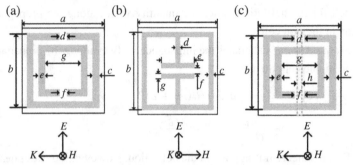

Figure 5.1 Three typical metamaterial unit cells used in the simulation. The excitations in the simulation are also given. (a) SRR structure. Substrate, FR4 with dielectric constant $\varepsilon = 4.4 + i0.001$ and thickness 0.25 mm, other parameters are $a = 2.5$ mm, $b = 2.2$ mm, $c = e = 0.2$ mm, $d = f = 0.22$ mm, and $g = 1.1$ mm. (b) ELC structure. Substrate, FR4 with dielectric constant $\varepsilon = 4.4 + i0.001$ and thickness about 0.2026 mm. Other parameters are $a = 3.333$ mm, $b = 3$ mm, $c = d = g = f = 0.2$ mm, and $e = 1.4$ mm. (c) SRR-wire structure. Substrate, FR4 with dielectric constant and thickness about 0.25 mm, $a = 2.5$ mm, $b = 2.2$ mm, $c = e = 0.2$ mm, $d = f = 0.3$ mm, $g = 1.1$ mm, and $h = 0.14$ mm.

Of course, the averaged quantities can be determined using the same equations. In Section 5.3, we show the correctness of these arguments.

5.3 Simulation Results for Typical Unit Cell Structures

In this section, we give a few typical examples to show the connection between the retrieved and averaged parameters. As the first example, we give the results of a SRR structure, with all parameters and excitations illustrated in Fig. 5.1(a).

With the retrieval method, effective permittivity and permeability for the SRR structure are obtained, as demonstrated in Fig. 5.2.

As we know, SRR is a typical magnetic resonator, which means the effective permeability will have a Lorentz-like response. This is clearly demonstrated in Fig. 5.2(b). However, the response should not be considered as a real Lorentz curve, since it is distorted by the spatial dispersion. The spatial dispersion also manifests itself in Fig. 5.2(a), where the retrieved permittivity is demonstrated and which should be a straight line since no electric resonance is involved. The response in Fig. 5.2(a) is called antiresonance in other references.

We have established the connection between the retrieved and averaged parameters using Eqs. (5.7) and (5.8). Hence it is quite easy to derive the corresponding parameters based on the data in Fig. 5.2. The results are demonstrated in Fig. 5.3; as clearly shown in the figure, the antiresonance phenomenon is eliminated in this case, and the magnetic response is totally a Lorenzian one.

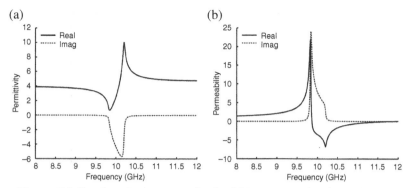

Figure 5.2 Retrieved parameters for the SRR structure in Fig. 5.1(a). (a) Permittivity. (b) Permeability.

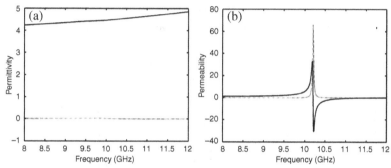

Figure 5.3 Parameters derived from the retrieved parameters in Fig. 5.2. (a) Permittivity. (b) Permeability. Please note that the solid line refers to the real part and the dotted line to the imaginary part.

Fig. 5.4 shows the fitted response using the standard Lorentz model. Then it is quite obvious that the derived permeability does follow the standard Lorentz response. The curve fit model and the corresponding fitted data are as follows:

$$\mu = \mu_\infty + \frac{(\mu_s - \mu_\infty)f_0^2}{f_0^2 - f^2 - if\gamma}, \tag{5.9}$$

where $\mu_\infty = 1.1225 + i0.0019$, $\mu_s = 1.3133 + i0.0027$, $f_0 = 10.2$ GHz, and $\gamma = 0.04$ GHz. The restoration of the "lost" Lorentz model has another advantage: one can get the frequency response of metamaterials by using only a few parameters. For the present case, only four coefficients are needed: μ_∞, μ_s, f_0, γ, and so on. And this leads to the rapid design for various metamaterials, since no time-consuming simulations are needed once these core parameters are obtained for a certain unit cell. To validate this argument, we recalculate the effective material parameters based on the curve-fitted results using Eqs. (5.7) and (5.8), and compare them with those obtained by directly using the retrieval process. The data are shown in Fig. 5.5.

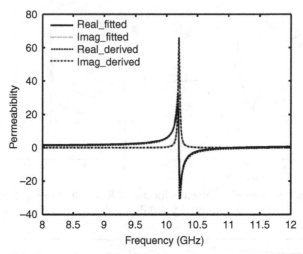

Figure 5.4 Comparison of the fitted Lorentz response and the derived permeability.

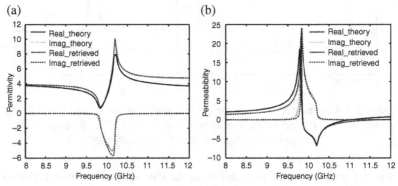

Figure 5.5 Parameters obtained from the theoretical calculation and from the S-parameter retrieval process, the SRR structure. (a) Permittivity. (b) Permeability.

Careful examination of Fig. 5.5 reveals that the material parameters obtained through theoretical calculation and those gotten by the retrieval process agree well with each other, and hence confirm the correctness of the theory.

Similar calculations are given for the electric resonator (i.e. the ELC structure and the SRR-wire structure), whose sizes and excitations are also given in Fig. 5.1. To be concise, only the results equivalent to Fig. 5.5 are shown, which are presented in Figs. 5.6 and 5.7, respectively. Corresponding coefficients and explanations are also provided in the following part.

The curve fit model and the corresponding fitted data for the ELC cell are as follows:

$$\varepsilon = \varepsilon_\infty + \frac{(\varepsilon_s - \varepsilon_\infty)f_0^2}{f_0^2 - f^2 - if\gamma}, \tag{5.10}$$

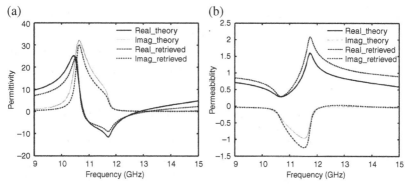

Figure 5.6 Parameters obtained from the theoretical calculation and from the S-parameter retrieval process, the ELC structure. (a) Permittivity. (b) Permeability.

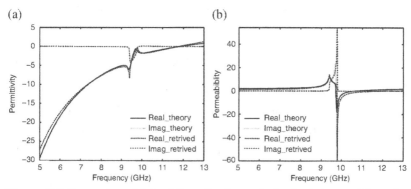

Figure 5.7 Parameters obtained from the theoretical calculation and from the S-parameter retrieval process, the SRR-wire structure. (a) Permittivity. (b) Permeability.

where $\varepsilon_\infty = 3.9366 + i0.1906, \varepsilon_s = 4.7246 + i0.2522, f_0 = 10.2\,\text{GHz}$, and $\gamma = 0.16\,\text{GHz}$.

And the curve fit model and the corresponding fitted data for the SRR-wire cell are as follows:

$$\varepsilon = \varepsilon_a(1 - \frac{f_0^2}{f^2}), \tag{5.11}$$

which is the Drude model and $\varepsilon_a = 5, f_0 = 11.8\,\text{GHz}$. The Lorentz model in Eq. (5.9) still holds for the permeability; however, the coefficients change to $\mu_\infty = 2.2911 - i0.0115, \mu_s = 2.9205 + i0.0068, f_0 = 9.4\,\text{GHz}$, and $\gamma = 0.02\,\text{GHz}$.

5.4 Summary

In this section we have shown a seemingly general effective medium theory for metamaterials. This theory can connect the two approaches previously mentioned (i.e., the retrieval method and the field averaging method). It also reveals the classical dielectric/permeability model hiding inside the effective material parameters and in the context of metamaterial. Moreover, the theory may be helpful for rapid metamaterial design due to the model's simplicity and accuracy.

6 Energy Densities in Artificial Metamaterials

In previous sections we obtained the effective parameters for various metamaterials. In this section we study another important problem concerning metamaterials: the energy density. Expressions for energy density in metamaterials are not explicitly defined; and the expression with $w_e = \frac{1}{2} D \cdot E$ in the electrostatic case and $w_m = \frac{1}{2} B \cdot H$ is obviously incorrect, since it will lead to negative energy density if one moves a little further, considering the negative permittivity and permeability in left-handed materials. Landau's classical formula,

$$\langle W \rangle = \frac{\varepsilon_0}{4} \frac{\partial [\omega \varepsilon(\omega)]}{\partial \omega} |E|^2 + \frac{\mu_0}{4} \frac{\partial [\omega \mu(\omega)]}{\partial \omega} |H|^2, \tag{6.1}$$

is not a general one either, since it deals with the time-averaged energy density of a harmonic electromagnetic wave, and the absorption of the medium is infinitesimal [50]. However, energy density has theoretical interest because the fundamental problem of energy propagation velocity related to causality and relativity is based on it.

Derivation of energy density in metamaterials, or dispersive and absorptive media in a more general sense, can be done in two ways; one is based on the electrodynamic approach [51–56], and the other on the equivalent circuit method [57, 58]. In recent years, researchers have also tried using the Lagrange formalism to address the problem, and satisfactory results have been obtained [59]. However, it should be noted that the stored energy may not be a correct or useful form in the quantization procedure. The problem is still a controversial one.

6.1 The Electrodynamic Approach

6.1.1 Lorentz Model Revisited

When using the electrodynamic approach, one has to know the effective permittivity and permeability of the artificial metamaterial. Based on knowledge from the previous sections, let's suppose that both the permittivity and the permeability have the Lorentz-type dispersion, that is, they can be expressed as follows:

$$\varepsilon_r = 1 + \frac{\omega_{ep}^2}{\omega_{e0}^2 - \omega^2 + i\gamma_e \omega}, \tag{6.2}$$

$$\mu_r = 1 + \frac{(F/\mu_0)\,\omega_{mp}^2}{\omega_{m0}^2 - \omega^2 + i\gamma_m\omega}. \tag{6.3}$$

Please note that Eqs. (6.2) and (6.3) are slightly different from those given in Section 2 for differentiation of the two material parameters, and F here is a constant. Please also note that Lorentz-type dispersion is chosen due to its generality, which covers the Drude-type dispersion by setting $\omega_{e0} = \omega_{m0} = 0$ and the cold plasma model by further setting $\gamma_e = \gamma_m = 0$.

As we have discussed in Section 2, the Lorentz-type dispersion is described by the following equation from the microscopic view,

$$m\ddot{r} = qE - \omega_{e0}^2 mr - \gamma_e m\dot{r}, \tag{6.4}$$

where the physical meaning for each variable is clearly explained in that section. Eq. (6.4) can be easily changed into the dynamic equation for the polarization vector considering $P = Nqr$, which reads

$$\dot{P} = \varepsilon_0\omega_{ep}^2 E - \omega_{e0}^2 P - \gamma_e \dot{P}. \tag{6.5}$$

And this is actually

$$\ddot{P} + \gamma_e\dot{P} + \omega_{e0}^2 P = \varepsilon_0\omega_{ep}^2 E, \tag{6.6}$$

where $\omega_{ep} = \frac{Nq^2}{\varepsilon_0 m}$ is used in Eqs. (6.4) and (6.5).

Similarly, the dynamic equation for the magnetization vector M can be obtained in the same way:

$$\ddot{M} + \gamma_m\dot{M} + \omega_{m0}^2 M = F\omega_{mp}^2 H. \tag{6.7}$$

6.1.2 Poynting Theorem inside the Metamaterial

Now that we have the dispersive medium at hand, it is not difficult to give Poynting's theorem in the time domain inside the medium [60]:

$$\nabla \cdot (E \times H) = -H \cdot \dot{B} - E \cdot \dot{D} - E \cdot J_s, \tag{6.8}$$

where J_s is the excitation electric current density. By integrating Eq. (6.8) over a volume V containing the source J_s, we get

$$-\int_V E \cdot J_s dV = \int_S (E \times H) \cdot dS + \int_V H \cdot \dot{B}dV + \int_V E \cdot \dot{D}dV, \tag{6.9}$$

where S is the closed surface surrounding volume V. Let's look at the integrand of the third term on the RHS of Eq. (6.9). We have

$$(E \cdot \dot{D}) = \varepsilon_0 E \cdot \dot{E} + E \cdot \dot{P} = \frac{\partial}{\partial t}\left(\frac{1}{2}\varepsilon_0|E|^2\right) + E \cdot \dot{P}, \tag{6.10}$$

whereas $D = \varepsilon_0 E + P$ is utilized. Similarly,

$$(H \cdot \dot{B}) = \mu_0 H \cdot \dot{H} + \mu_0 H \cdot \dot{M} = \frac{\partial}{\partial t}\left(\frac{1}{2}\mu_0 |H|^2\right) + \mu_0 H \cdot \dot{M}, \quad (6.11)$$

and $B = \mu_0 H + \mu_0 M$ is applied likewise.

Based on the dynamic equation, Eq. (6.6), and replacing the second E on the RHS of Eq. (6.10), we easily obtain

$$
\begin{aligned}
(E \cdot \dot{D}) &= \frac{\partial}{\partial t}\left(\frac{1}{2}\varepsilon_0 |E|^2\right) + \frac{1}{\varepsilon_0 \omega_{ep}^2}\left(\ddot{P} + \gamma_e \dot{P} + \omega_{e0}^2 P\right) \cdot \dot{P} \\
&= \frac{\partial}{\partial t}\left(\frac{1}{2}\varepsilon_0 |E|^2\right) + \frac{1}{2\varepsilon_0 \omega_{ep}^2}\frac{\partial}{\partial t}\left(|\dot{P}|^2 + \omega_{e0}^2 |P|^2\right) + \frac{\gamma_e}{\varepsilon_0 \omega_{ep}^2}|\dot{P}|^2.
\end{aligned}
$$
$$(6.12)$$

Please note that bold letters are used for vectors during the substitution due to the isotropy of the materials considered. Similarly,

$$
\begin{aligned}
(H \cdot \dot{B}) &= \frac{\partial}{\partial t}\left(\frac{1}{2}\mu_0 |H|^2\right) + \frac{1}{F\omega_{mp}^2}\left(\ddot{M} + \gamma_m \dot{M} + \omega_{m0}^2 M\right) \cdot \dot{M} \\
&= \frac{\partial}{\partial t}\left(\frac{1}{2}\mu_0 |H|^2\right) + \frac{\mu_0}{2F\omega_{mp}^2}\frac{\partial}{\partial t}\left(|\dot{M}|^2 + \omega_{m0}^2 |M|^2\right) + \frac{\mu_0 \gamma_m}{F\omega_{mp}^2}|\dot{M}|^2.
\end{aligned}
$$
$$(6.13)$$

We then substitute Eqs. (6.10) and (6.13) into Eq. (6.9),

$$-\int_V E \cdot J_s dV = \int_S (E \times H) \cdot dS + \dot{W} + \int_V \frac{\mu_0 \gamma_m}{F\omega_{mp}^2}|\dot{M}|^2 dV + \int_V \frac{\gamma_e}{\varepsilon_0 \omega_{ep}^2}|\dot{P}|^2 dV,$$
$$(6.14)$$

where $W = W_e + W_m$ represents the total energy stored inside the volume V, and $W_{e,m} = \int_V w_{e,m} dV$ is the corresponding electric/magnetic energy inside V, respectively. Obviously, w_e and w_m can be defined as electric and magnetic energy density, which are

$$w_e = \left(\frac{1}{2}\varepsilon_0 |E|^2\right) + \frac{1}{2\varepsilon_0 \omega_{ep}^2}\left(|\dot{P}|^2 + \omega_{e0}^2 |P|^2\right), \quad (6.15)$$

$$w_m = \left(\frac{1}{2}\mu_0 |H|^2\right) + \frac{\mu_0}{2F\omega_{mp}^2}\left(|\dot{M}|^2 + \omega_{m0}^2 |M|^2\right). \quad (6.16)$$

Equation (6.16) implies a conservation of energy, which is

$$P_s = P + \left(\dot{W}_e + \dot{W}_m\right) + P_{e\sigma} + P_{m\sigma}, \quad (6.17)$$

where $P = \int_S (E \times H) \cdot dS$ is the total power flowing outwards through the surface, and $P_s = -\int_V E \cdot J_s dV$ is the total power provided by the source current inside V.

Actually, the last two terms in Eqs. (6.14) and (6.17) represent the heat energy due to the loss, which can be expressed as

$$P_{e\sigma} = \frac{\gamma_e}{\varepsilon_0 \omega_{ep}^2} \int_V dV |\frac{\partial \mathbf{P}}{\partial t}|^2, \tag{6.18}$$

$$P_{m\sigma} = \frac{\mu_0 \gamma_m}{F \omega_{mp}^2} \int_V dV |\frac{\partial \mathbf{M}}{\partial t}|^2. \tag{6.19}$$

Then, Eq. (6.17) states that the power provided by the source is equal to the sum of the power flowing outwards through the surface, the heat produced inside the medium per second, and the power stored inside the volume.

Equations (6.8) through (6.19) apply to arbitrary time-domain electromagnetic waves propagating inside the metamaterials, with permittivity and permeability given by Eqs. (6.2) and (6.3).

6.1.3 Energy Density inside the Metamaterial with Lorentz-Type Dispersion

With Eqs. (6.8) through (6.19) at hand, we can calculate the energy density for various materials and arbitrary time-domain waves. For the time-harmonic case, one has, for the time-averaged energy density, the following expressions:

$$\overline{w}_e = \left(\frac{1}{4}\varepsilon_0 |\mathbf{E}|^2\right) + \frac{1}{4\varepsilon_0 \omega_{ep}^2} \left(\omega^2 + \omega_{e0}^2\right) |\mathbf{P}|^2, \tag{6.20}$$

$$\overline{w}_m = \left(\frac{1}{4}\mu_0 |\mathbf{H}|^2\right) + \frac{\mu_0}{4F\omega_{mp}^2} \left(\omega^2 + \omega_{m0}^2\right) |\mathbf{M}|^2. \tag{6.21}$$

Utilizing the dynamic equations, Eqs. (6.6) and (6.7), and considering the harmonic situation, we obtain

$$\overline{w}_e = \frac{1}{4}\varepsilon_0 \left(1 + \frac{\left(\omega^2 + \omega_{e0}^2\right)\omega_{ep}^2}{\left(\omega_{e0}^2 - \omega^2\right)^2 + \gamma_e^2\omega^2}\right) |\mathbf{E}|^2, \tag{6.22}$$

$$\overline{w}_m = \frac{1}{4}\mu_0 \left(1 + \frac{\left(\omega^2 + \omega_{m0}^2\right)F\omega_{mp}^2}{\left(\omega_{m0}^2 - \omega^2\right)^2 + \gamma_m^2\omega^2}\right) |\mathbf{H}|^2. \tag{6.23}$$

6.2 The Equivalent Circuit Approach

The circuit-based method can also be used to get the energy density expression inside dispersive and absorptive materials. This approach involves, first, identification of the circuit model of the electromagnetic problem, establishment of equivalence between corresponding parameters, then calculation of the energy stored inside the circuit, and finally the transformation from the circuit parameters into the field quantities. In this section, we show in the time-harmonic case how the energy density is found inside a Lorentz-type medium.

6.2.1 Tretyakov's Method

As before, let's suppose that the permittivity of the material has the form shown in Eq. (6.2). To get the circuit equivalence, we further imagine that the material is filled inside a parallel-plate capacitor $C = \frac{\varepsilon S}{d}$, whose plate area is S and the distance betweeen the plates is d. Then, in the time-harmonic case, this imaginary capacitor has the following input admittance:

$$Y = i\omega C = i\omega \frac{\varepsilon_0 S}{d} \left(1 + \frac{\omega_{ep}^2}{\omega_{e0}^2 - \omega^2 + i\gamma_e \omega} \right)$$

$$= i\omega C_0 \left(1 + \frac{\omega_{ep}^2}{\omega_{e0}^2 - \omega^2 + i\gamma_e \omega} \right). \tag{6.24}$$

This expression can be easily changed into the following form:

$$Y = i\omega C_0 + \cfrac{1}{\cfrac{\omega_{e0}^2}{i\omega C_0 \omega_{ep}^2} + \cfrac{i\omega}{C_0 \omega_{ep}^2} + \cfrac{\gamma_e}{C_0 \omega_{ep}^2}} = i\omega C_0 + \cfrac{1}{\cfrac{1}{i\omega C_e} + i\omega L_e + R_e}, \tag{6.25}$$

where

$$C_e = \frac{C_0 \omega_{ep}^2}{\omega_{e0}^2}, L_e = \frac{1}{C_0 \omega_{ep}^2}, R_e = \frac{\gamma_e}{C_0 \omega_{ep}^2}. \tag{6.26}$$

Then, it is easy to identify the equivalent circuit of the "capacitor," which is depicted in Fig. 6.1. The circuit has two branches: one is a capacitor C_0, and the other is a series connection by resistor R_e, inductor L_e, and capacitor C_e.

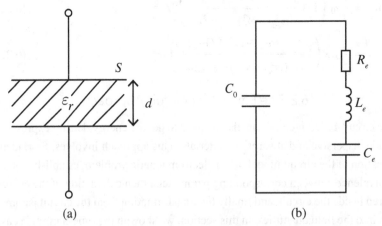

Figure 6.1 The imaginary capacitor (a) and its equivalent circuit (b).

Thus, the time-averaged stored reactive energy (heat is excluded) inside the circuit is

$$W = \frac{1}{4}\left(C_0|V_{c0}|^2 + L_e|I_{Le}|^2 + C_e|V_{ce}|^2\right).$$ (6.27)

Using circuit theory, we have

$$I_{Le} = \frac{V_{c0}}{R_e + i\omega L_e + \frac{1}{i\omega C_e}}, V_{ce} = \frac{I_{Le}}{i\omega C_e} = \frac{1}{i\omega C_e}\frac{V_{c0}}{R_e + i\omega L_e + \frac{1}{i\omega C_e}}.$$ (6.28)

On the other hand, this energy can be expressed using time-averaged electric energy density, which is

$$W = \overline{w}_e S d.$$ (6.29)

Moreover, $V_{c0} = Ed$ for the capacitor in Fig. 6.1, where is the electric field inside it.

Hence the equality of Eqs. (6.27) and (6.29) gives the following expression:

$$\overline{w}_e = \frac{1}{4}\varepsilon_0\left(1 + \frac{\left(\omega^2 + \omega_{e0}^2\right)\omega_{ep}^2}{\left(\omega_{e0}^2 - \omega^2\right)^2 + \gamma_e^2\omega^2}\right)|E|^2.$$ (6.30)

Please note that the coefficient is $\frac{1}{2}$ in the original paper [58]; this is because effective quantities are used for the phasors in the expression, while in our expression, the maximum value is used. As can be verified, the electric energy density agrees perfectly with that given in Eq. (6.22).

If one wants to deal with the magnetic energy density in a Lorentz-type material, similar methods apply. However, one has to find an imaginary "inductor" filled with the specific material and get the equivalent circuit accordingly.

Though the circuit-based method is quite simple and straightforward, care should be taken to identify the equivalent circuit model. As mentioned by Tretyakov [58], since different circuits can have the same input admittance, the circuit model is not unique. One has to check the physical meaning of the circuit and the structure of the metamaterial, so they are truly equivalent!

6.2.2 Fung's Method

Fung also proposed a circuit method in 1978 [57]. His method begins with the observation that the work done per unit volume in setting up the field in a medium is

$$w_e = \int_t E \cdot \dot{D} dt.$$ (6.31)

However, in the circuit case, for a circuit element with an impedance $Z(\omega)$, the work done on this element is

$$W = \int_t V I dt = \int_t V \dot{q} dt. \tag{6.32}$$

Comparison between the two equations leads to the following equivalence (considering the isotropy of the material and omitting the vector property in Eq. (6.31)):

$$V \leftrightarrow E, q \leftrightarrow D. \tag{6.33}$$

In the time-harmonic case, using the phasor representation,

$$D = \varepsilon(\omega)E, \tag{6.34}$$

and

$$q = \frac{I}{i\omega} = \frac{V}{i\omega Z(\omega)}, \tag{6.35}$$

we then find another identity:

$$\frac{1}{i\omega Z(\omega)} \leftrightarrow \varepsilon(\omega). \tag{6.36}$$

In this regard, a medium with permittivity profile $\varepsilon(\omega)$ can be considered as a circuit element having the impedance $Z(\omega)$, using Eq. (6.36).

For the material with Lorentz-type dispersion, as given in Eq. (6.2), the identity in Eq. (6.36) leads to the following equation:

$$\frac{1}{Z(\omega)} = i\omega\varepsilon_0 + \frac{i\omega\varepsilon_0\omega_{ep}^2}{\omega_{e0}^2 - \omega^2 + i\gamma_e\omega}. \tag{6.37}$$

Actually, Eq. (6.37) is essentially the same as Eq. (6.25), where C_0 is replaced by ε_0. As a result, the same equivalent circuit is obtained, as shown in Fig. 6.1.

Again, Eq. (6.27) can be used to get the reactive energy in the circuit. Noticing the equivalence of Eqs. (6.33), (6.36), and the identity between C_0 and ε_0, Eq. (6.30) can be reproduced. We thus arrive at the same end through different means!

6.3 Summary

Energy density is a very controversial problem in the metamaterials sector. In this section, we use two techniques to derive this quantity for different metamaterials, and both lead to the same expressions. The method can be applied to other complex metamaterials too. As we mentioned in the earlier part of the section, other approaches exist for the energy density expression in metamaterials (e.g. the Lagrange formalism [59]). The reader is strongly encouraged to explore the reference works at the end of the Element for further reading.

7 Effective Medium Theory for 2D Metasurfaces

A metasurface is a kind of two-dimensional (2D) metamaterial, and the methods of analysis and design are different. In this section we introduce the effective parameters and design methods of metasurfaces, including surface refractive index and surface impedance. This is the theoretical basis of functional devices and other applications based on metasurfaces.

A metasurface is composed of a series of arrayed periodic or aperiodic elements in two dimensions. For three-dimensional (3D) metamaterials, the EM properties are described by the effective electric permittivity and magnetic permeability. However, they are not suitable for 2D metasurfaces. Some analysis methods for metasurfaces have been proposed, such as the generalized surface transmission condition [61–63], transverse resonance method [64–67], and so on.

The method based on the generalized surface transmission condition (GSTC) [61–63] is to construct a complete mathematical model to compute the reflection and transmission coefficients of metasurfaces by introducing electric and magnetic polarization intensities in Maxwell's equations and classical EM boundary conditions.

Assume that the propagation direction of the EM wave is along the longitudinal direction of the metasurface. For the transverse direction, on the interface of the metasurface and the outside medium (air in the general case), the sum of the admittances along the positive normal of the interface and the negative normal of the interface is zero, which is the transverse resonance condition [64–67]. The analysis method for metasurfaces based on the transverse resonance condition is called the transverse resonance method. Generally, this method is used for the modeling analysis of surface waves.

The design methods of metasurfaces based on the classical electromagnetism theory mainly include Babinet's complementary principle [68], the generalized Snell's law [69], and the method of Huygens' surface [70]. All these methods introduce the basic physical laws in electromagnetism to the theoretical analysis of metasurface. In most cases, these three methods are used for the design and analysis of spatial waves [68–70].

Based on the metasurface, some functional devices have been designed to control the amplitude, phase, polarization, and power flow of EM waves, which can even be extended to system-level applications. In Sections 7.1 through 7.4,

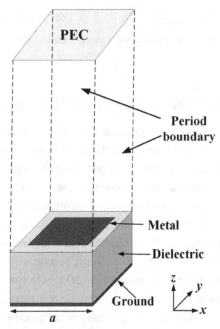

Figure 7.1 Eigenmode simulation for retrieving the surface refraction index.

we discuss the effective parameters and some design methods and applications of the EM metasurfaces.

7.1 Effective Surface Refractive Index

In most cases, the grounded dielectric structures support transverse-magnetic (TM) mode surface waves, while the ungrounded dielectric structures support transverse-electric (TE) mode surface waves [71]. Here, we choose a grounded patch structure, as shown in Fig. 7.1. The structure is isotropic, and we first retrieve the dispersion relation with the help of the numerical method. The setup of the element for eigenmode analysis is shown in Fig. 7.1 with surrounded periodic boundary conditions and PEC boundaries up and down. However, beyond a certain distance from the surface, due to the attenuation of surface waves, the top PEC boundary can be approximately considered as the open boundary.

For the grounded structure, the lower boundary is naturally a PEC boundary condition. Based on the eigenmode simulation setup, we can get the dispersion relation for the structure, as shown in Fig. 7.2, which demonstrates the relationship between the phase shift and the corresponding frequency. It should be noted that the phase shift reflects the surface refraction index directly, which is similar to the relationship between the corresponding physical quantities

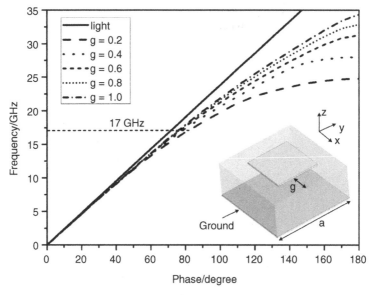

Figure 7.2 Dispersion curves of unit structure with different sizes.

for space waves. Based on the dispersion relation, we can retrieve the surface refraction index of metasurface structures. Fig. 7.2 also illustrates how the surface refraction index can be retrieved at 17 GHz. First, we can obtain the phase shifts ϕ from Fig. 7.2 for different sizes (the gap between patch structure g) of the structure at 17 GHz, and the corresponding phases satisfy $\phi = k_t a$, where k_t is the wave number of surface waves. We can then calculate the surface refraction index of the element at different sizes according to the following relationship:

$$n = c/v_t = k_t c/\omega_t = \phi c/a\omega_t, \tag{7.1}$$

in which c is the velocity of light in free space, and ω_t is angle frequency at 17 GHz. Equation (7.1) is useful for the design of surface devices. Note that the leaky and ohmic losses of the grounded structure are quite tiny and ignored.

7.2 Surface Impedance Characterization

The transverse resonance technique is usually used for surface waves because the wave propagation direction is along the interface. By using the transverse resonance method, we can obtain the dispersion equation of the element for metasurfaces [64].

We first discuss the theoretical modeling of the isotropic elements of the metasurface based on the transverse resonance method [64]. To construct the transverse resonance conditions for solving the dispersion equation, we first need the impedance η_{sheet} of the sheet attached above the dielectrics. Then we

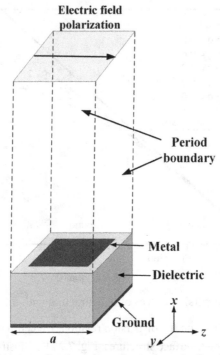

Figure 7.3 The setup for sheet impedance (η_{sheet}) of an isotropic patch.

can calculate the impedance by using the numerical method, and the simulation model is shown in Fig. 7.3.

Assuming that the y-polarized or z-polarized waves are incident on the element of the metasurface vertically and propagate along the $-x$ direction, we first calculate the input impedance η_{in} of the spatial wave above the sheet by using the numerical method. Then we build the relationship, as shown in Eq. (7.2), between the impedance of the sheet and the input impedance by using the transmission line model shown in Fig. 7.4,

$$\frac{1}{\eta_{sheet}} = \frac{1}{\eta_{in}} - \frac{1}{-i\eta_1 \tan(k_1 d)},$$ (7.2)

in which η_1 and k_1 are the wave impedance and wave number of propagating waves in dielectrics.

After that, the surface of the metal is considered as the interface for the analysis in the transverse resonance method. Using the transverse resonance condition, that is, the sum of the admittances along the positive normal of the surface and the negative normal of the interface is zero, we reach the following relation:

$$\frac{1}{\eta_{up}} + \frac{1}{\eta_{down}} = 0.$$ (7.3)

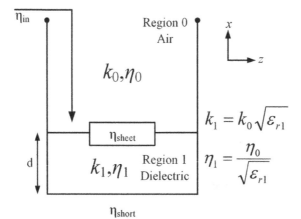

Figure 7.4 Effective transmission-line model when the EM waves are incident vertically [64]. η_0 and k_0 are wave impedance and wave number in free space, respectively. ε_{r1} is relative permittivity in dielectrics.

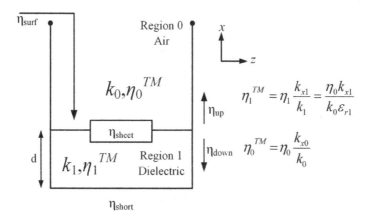

Figure 7.5 The modified transmission line model for calculation of surface impedance [64].

Similarly, we can set up the dispersion equation for the surface wave based on the transverse resonance method analysis using the surface-wave-based transmission line model, as shown in Fig. 7.5:

$$\frac{1}{\eta_{surf}} = \frac{1}{\eta_{sheet}} + \frac{1}{-i\frac{\eta_0 k_{x1}}{k_0 \varepsilon_{r1}} \tan(k_{x1} d)}, \tag{7.4}$$

where

$$k_{x1} = \sqrt{k_0^2(\varepsilon_{r1} - 1) + \left(\frac{\eta_{surf} k_0}{\eta_0}\right)^2} \tag{7.5}$$

is the transverse wave number of the surface wave in the medium, and η_{surf} is the surface impedance of the surface wave. The structure shown in Fig.

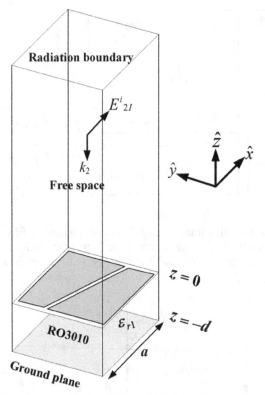

Figure 7.6 The anisotropic elements analysis by using transverse resonance method [65].

7.3 is the grounded structure, which supports the TM mode surface waves [71]. Hence, we construct the surface-wave dispersion equation of the structure according to Eqs. (7.4) and (7.5), from which we can get the values of the surface refractive index and the impedance at any frequency.

7.3 Tensor Surface Impedance

The transverse resonance method is also used for anisotropic elements of the metasurface [65], in which the basic idea and the method for obtaining the dispersion equation are similar to those for the isotropic cases, but the analysis process is more complicated.

Next, we discuss the analysis of anisotropic elements of the metasurface [65] by using the transverse resonance method. As shown in Fig. 7.6, there is an oblique slot in the patch, which generates the anisotropy. For the anisotropic metasurface element, the mixed surface waves of TM and TE modes are supported instead of the pure TM mode or pure TE mode. Furthermore, the sheet impedance of the element is a second-order tensor instead of a scalar, namely,

$$\overline{\overline{\eta}}_{\text{sheet}} = \begin{bmatrix} \eta^s_{xx} & \eta^s_{xy} \\ \eta^s_{yx} & \eta^s_{yy} \end{bmatrix}. \tag{7.6}$$

It establishes the relationship between the surface electric field and the surface current (magnetic field) at $z = 0$ as follows:

$$\begin{bmatrix} E_x \\ E_y \end{bmatrix} = \begin{bmatrix} \eta^s_{xx} & \eta^s_{xy} \\ \eta^s_{yx} & \eta^s_{yy} \end{bmatrix} \begin{bmatrix} J_x \\ J_y \end{bmatrix}. \tag{7.7}$$

Similar to the calculation of surface impedance for the isotropic element, the calculation of the impedance tensor for the anisotropic element needs two simulations: one for the x-polarized incident waves on the metasurface element, and the other for the y-polarized waves, respectively. After that, we can obtain the input impedance matrix by using the boundary conditions on the interface between the medium (Region I) and free space (Region II). For convenience, we use the admittance matrix Y^{in} instead of the impedance matrix η^{in}, and the input admittance matrix for each matrix element is expressed as

$$Y^{in}_{xx} = \frac{E^I_{y2} H^{II}_{y2} - E^{II}_{y2} H^I_{y2}}{E^I_{x2} E^{II}_{y2} - E^{II}_{x2} E^I_{y2}}, \tag{7.8a}$$

$$Y^{in}_{xy} = \frac{E^{II}_{x2} H^I_{y2} - E^I_{x2} H^{II}_{y2}}{E^I_{x2} E^{II}_{y2} - E^{II}_{x2} E^I_{y2}}, \tag{7.8b}$$

$$Y^{in}_{yx} = \frac{E^{II}_{y2} H^I_{x2} - E^I_{y2} H^{II}_{x2}}{E^I_{x2} E^{II}_{y2} - E^{II}_{x2} E^I_{y2}}, \tag{7.8c}$$

$$Y^{in}_{yy} = \frac{E^I_{x2} H^{II}_{x2} - E^{II}_{x2} H^I_{x2}}{E^I_{x2} E^{II}_{y2} - E^{II}_{x2} E^I_{y2}}, \tag{7.8d}$$

in which I and II denote the x-polarized and y-polarized incident waves, and E_{x2}, E_{y2}, H_{x2}, and H_{y2} denote the x component and y component of the total electric and magnetic fields. According to the transmission-line model, we get

$$\begin{bmatrix} Y^{in}_{xx} & Y^{in}_{xy} \\ Y^{in}_{yx} & Y^{in}_{yy} \end{bmatrix} = \begin{bmatrix} Y^s_{xx} + \frac{\cot(k_1 d)}{j\eta_1} & Y^s_{xy} \\ Y^s_{yx} & Y^{in}_{xx} + \frac{\cot(k_1 d)}{i\eta_1} \end{bmatrix}$$

$$= \overline{\overline{Y}}_{\text{sheet}} + \begin{bmatrix} \frac{1}{-i\eta_1 \tan(k_1 d)} & 0 \\ 0 & \frac{1}{-i\eta_1 \tan(k_1 d)} \end{bmatrix}. \tag{7.9}$$

From Eqs. (7.8) and (7.9), we can obtain the admittance matrix $\overline{\overline{Y}}_{\text{sheet}}$ if we know the input admittance matrix. Finally, we can achieve the equivalent surface admittance of the anisotropic elements after some algebra calculations [66],

$$
\bar{\bar{Y}}_{\text{surf}} = \begin{bmatrix} Y_{xx}(\theta) & Y_{xy}(\theta) \\ Y_{yx}(\theta) & Y_{yy}(\theta) \end{bmatrix}
$$

$$
= \begin{bmatrix} Y^s_{xx} & Y^s_{xy} \\ Y^s_{yx} & Y^s_{yy} \end{bmatrix} + R^T(-\theta) \begin{bmatrix} i(Y_1 \frac{k_1}{k_{z1}}) \cot(k_{z1}d) & 0 \\ 0 & i(Y_1 \frac{k_{z1}}{k_1}) \cot(k_{z1}d) \end{bmatrix} R(-\theta)
$$

$$(7.10)$$

in which

$$
R(\theta) = \begin{bmatrix} \cos(\theta) & \sin(-\theta) \\ \sin(\theta) & \cos(\theta) \end{bmatrix}
\tag{7.11}
$$

is the rotation matrix and θ is the angle between the propagation direction of the surface wave on the metasurface and the x axis, and

$$
\bar{\bar{Y}}_{\text{sheet}} = \begin{bmatrix} Y^s_{xx} & Y^s_{xy} \\ Y^s_{yx} & Y^s_{yy} \end{bmatrix}
\tag{7.12}
$$

can be calculated by Eq. (7.9), where $Y_1 = \sqrt{\varepsilon_1/\mu_1}$, ε_1 and μ_1 are the permittivity and permeability of the dielectrics, and k_{z1} is the transverse wave number of the surface wave in the dielectrics, respectively.

Because the element is anisotropic, the surface admittance values are different when the surface waves propagate on the surface along different directions, and the specific values can be calculated by Eq. (7.10). We can understand Eq. (7.10) as follows: the admittance tensor of surface waves is obtained by adding the transmission-line admittance tensor with the rotation angle $-\theta$ to matrix $\bar{\bar{Y}}_{\text{sheet}}$. So far, Eq. (7.10) can be used to calculate the effective surface admittance or impedance on the anisotropic metasurface when the surface waves propagate along any direction.

7.4 Design Methods and Applications of Metasurface-Based Devices

In this section we discuss the design method and applications of metasurfaces based on the effective parameter representation.

7.4.1 Generalized Snell's Law

According to Snell's law, when plane waves are incident to the interface between two kinds of homogeneous media, the reflection angle is equal to the incident angle, and the refraction angle is calculated by

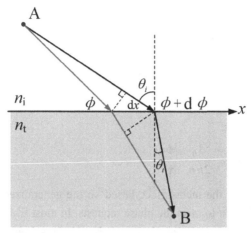

Figure 7.7 The illustration for generalized Snell's refraction law [69].

$$\frac{\sin(\theta_t)}{\sin(\theta_i)} = \frac{n_i}{n_t}, \tag{7.13}$$

in which θ_i and θ_t are the incident and refraction angles, respectively, and n_i and n_t are the refractive indexes of materials for the incident wave and the refractive wave, respectively.

In the traditional Snell's law, the phase is continuously changing on the interface of two kinds of homogeneous media, and it has been extended to the generalized Snell's law by introducing discontinuous phase changes along the tangential direction of the interface [69]. We will discuss the refraction characteristics based on the generalized Snell's law.

As shown in Fig. 7.7, a plane wave with incident angle θ_i illuminates the interface of two kinds of media, and there is phase variation $d\Phi$ at the interface along the x direction. Assuming that the two wave paths are very close to the actual optical paths, because the total phase differences of the waves in two paths are the same (according to Fermat's principle, the actual light path is a stationary path), we have

$$[k_0 n_i \sin(\theta_i) dx + (\Phi + d\Phi)] - [k_0 n_t \sin(\theta_t) dx + \Phi] = 0, \tag{7.14}$$

where $k_0 = 2\pi/\lambda_0$, and dx is the distance of the incident points on the interface of the two paths. According to Eq. (7.14), we obtain

$$n_t \sin(\theta_t) - n_i \sin(\theta_i) = \frac{\lambda_0}{2\pi} \frac{d\Phi}{dx}. \tag{7.15}$$

This formula is the generalized Snell's law of refraction obtained by introducing the phase discontinuity. Similarly, we can get the generalized Snell's law of reflection by introducing the phase discontinuity:

| 0° | 45° | 90° | 135° | 180° | 225° | 270° | 315° |

Figure 7.8 The V-shaped structures to realize 360° cross-polarized wave phase control.

$$\sin(\theta_r) - \sin(\theta_i) = \frac{\lambda_0}{2\pi n_i} \frac{d\Phi}{dx}. \tag{7.16}$$

The key point of the metasurface based on the generalized Snell's law is to design the element to meet the phase requests. In most EM device designs, especially in the wavefront-control devices, the elements should cover all phase regulation ability in 360°. In Ref. [69], a kind of single-layer, V-shaped metasurface element was put forward, as shown in Fig. 7.8. By changing the open size and open direction of the V-shaped structures, one can realize 360° phase control for the cross-polarized waves.

The basic structures shown in Fig. 7.8 can greatly satisfy the design requirements of the wave phase control. Some work has been reported, such as the beam bending experiment [72], the generation of interference patterns in holographic imaging [73], and so on.

7.4.2 Huygens' Surface

From the view of Huygens' principle, each point on the wave front is the secondary source of the next wave [71]. Then, Love and Schelkunoff developed the theory and formulated the equivalence principle [71]. Maxwell's equations with electric and magnetic currents J and M are expressed as

$$\nabla \times H = J + \frac{\partial D}{\partial t}, \tag{7.17a}$$

$$\nabla \times E = -M - \frac{\partial B}{\partial t}. \tag{7.17b}$$

As shown in Fig. 7.9, the excited source is set in Region I, and the fields in this area are represented as E_1 and H_1. In Region II, the desired fields are expressed as E_2 and H_2, which can be totally different from the fields in Region I. According to the equivalence principle, the equivalent electric and magnetic currents on the border (dotted line shown in Fig. 7.9) can be expressed as

$$J_s = \hat{n} \times (H_2 - H_1), \tag{7.18a}$$

$$M_s = -\hat{n} \times (E_2 - E_1), \tag{7.18b}$$

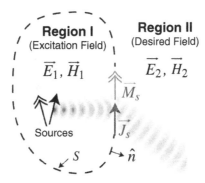

Figure 7.9 The illustration of equivalence principle [70].

in which $\hat{n}$ is the outside normal vector of the boundary. From Eqs. (7.17) and (7.18), we can control the wave reflection and transmission on the interface if we can control the equivalent electric and magnetic currents. Hence, it provides a strong tool to design metasurfaces.

From the metasurface model based on the GSTC principle [61–63, 70], we have

$$J_s = -i\omega\bar{\bar{\alpha}}_e^{\text{eff}}\cdot E_{t,av}\big|_s, \tag{7.19a}$$

$$M_s = -i\omega\bar{\bar{\alpha}}_m^{\text{eff}}\cdot H_{t,av}\big|_s, \tag{7.19b}$$

in which $\bar{\bar{\alpha}}_e^{\text{eff}}$ and $\bar{\bar{\alpha}}_m^{\text{eff}}$ are the effective surface polarization, and the corresponding electric admittance and magnetic impedance can be defined as $\bar{\bar{Y}}_{es} = -i\omega\bar{\bar{\alpha}}_e^{\text{eff}}$ and $\bar{\bar{Z}}_{ms} = -i\omega\bar{\bar{\alpha}}_m^{\text{eff}}$, respectively. If the wave is incident along the x direction, and the metasurface is put on the yoz plane and is isotropic, then $Y_{es} = Y_{es}^{yy} = Y_{es}^{zz}$ and $Z_{ms} = Z_{ms}^{yy} = Z_{ms}^{zz}$. Thus, by using the formulas of reflectivity and transmissivity of metasurface for a vertically incident wave [70] and Eq. (7.18), we have

$$Y_{es} = \frac{2(1 - T - R)}{\eta(1 + T + R)}, \tag{7.20a}$$

$$Z_{ms} = \frac{2\eta(1 - T + R)}{(1 + T - R)}, \tag{7.20b}$$

where $\eta = \sqrt{\mu/\varepsilon}$ is wave impedance in free space, and T and R are the transmission coefficient and reflection coefficient on the interface, respectively. If the normalized electric admittance and magnetic impedance are equal, that is $Y_{es}\eta = Z_{ms}/\eta$, the reflection coefficient R on the interface will be zero by Eq. (7.20). That is to say, the interface implements the total transmission control without any reflection for the incident EM wave. Hence, we can realize the wave transmission with 100% transmittance and 360° of all phase controls by adjusting the amplitude of the electric admittance and magnetic impedance.

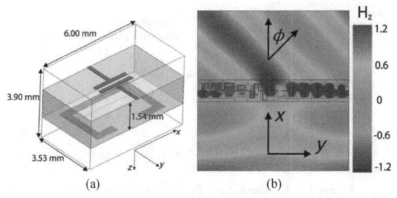

Figure 7.10 (a) The element of Huygens' metasurface, (b) beam deflection without any reflection based on Huygens' metasurface [70].

An element of the Huygens' metasurface [70] is shown in Fig. 7.10 (a). The upper and lower surfaces are composed of double-sided, printed metal structure, and the middle is a dielectric. The upper structure can be used to control the electric resonance, and the lower structure can be used to control the magnetic resonance. Hence, the electric and magnetic resonances can be designed at the same time, and EM waves can transmit through the interface without reflections, while the phase can be adjustable with 360°. From the phase discontinuity in the generalized Snell's law, we can easily realize the deflection transmission of incident beams with almost no reflection too, which is illustrated in Fig. 7.10(b).

7.4.3 Holography Metasurfaces

In this subsection we introduce the concept of holographic impedance surface, which can also be regarded as a special metasurface. The concept of holography was first proposed in the field of optics, and the applications mainly focus on optical imaging. However, the technology has also been introduced in the microwave field, and the applications mainly focus on the antenna design [74]. In general, the holographic antenna design can be divided into two steps: first of all, to generate the interference record in the hologram between the reference wave (exciting source) and the objective wave the needed radiation pattern. The mathematical expression is as follows:

$$\Psi = |\psi_{\text{ref}} + \psi_{\text{obj}}|, \tag{7.21}$$

where ψ_{ref} refers to reference wave, and ψ_{obj} means objective wave, which is the needed radiation pattern. We can extend Eq. (7.20) to the following one (the intensity of the interfered waves):

$$|\Psi|^2 = |\psi_{\text{ref}} + \psi_{\text{obj}}|^2 = |\psi_{\text{ref}}|^2 + |\psi_{\text{obj}}|^2 + \psi_{\text{ref}}^*\psi_{\text{obj}} + \psi_{\text{obj}}^*\psi_{\text{ref}}, \tag{7.22}$$

in which the superscript "*" denotes the conjugation of a complex variable, and $|\Psi|^2$ means the strength of the holographic interferograms. According to the holographic reproduction theory, if we use the same reference wave ψ_{ref} to excite the interference pattern again, we can get the needed wave, and its mathematical expression is $|\psi_{ref}|^2\psi_{obj}$.

After the establishment of the mathematical expression of the holographic interference pattern, the subsequent question is how to characterize the interference pattern by using one proper physical quantity. The answer is shown in Ref. [75], in which Fong et al. initiatively introduce surface impedance to characterize holographic interference patterns. Note that the surface impedance here refers to the surface wave impedance, and as we know from Section 7.3, the artificial EM structure with a ground plate supports the TM-mode surface waves, and the structure without a ground plate supports the TE-mode surface waves. In Ref. [75], the authors use the patch structures with ground to support the TM-mode surface waves. Herein, we focus on the TM-mode surface waves; and for the TE-mode surface waves, the EM duality principle can be used to analyze the case.

Surface wave impedance refers to the ratio of the tangential component of the electric field and the tangential component of the magnetic field on the metasurface. For each element of the metasurface, if the normal of the element is the z direction, the surface impedance can be defined as

$$Z_s = \int_{cell} \frac{E_x}{H_y} ds, \tag{7.23}$$

where E_x and H_y are the tangential electric field and the magnetic field on the element of the impedance surface, respectively, and ds is the differential area in the elements. This definition is straightforward but is inconvenient for the extraction and calculation of the surface impedance. Generally, the surface impedance is defined as [76]

$$Z_s = -iZ_0\frac{k_z}{k}, \tag{7.24}$$

in which i is the imaginary unit, Z_0 is the wave impedance in free space, and k_z and k are the normal wave number and the total wave number of the surface wave, respectively. And the wave number equation of the surface wave is

$$k^2 = k_z^2 + k_t^2, \tag{7.25}$$

in which k_t is the transverse wave number. Insert Eq. (7.24) into Eq. (7.23) and we obtain

$$Z_s = iZ_0\sqrt{n^2 - 1}, \tag{7.26}$$

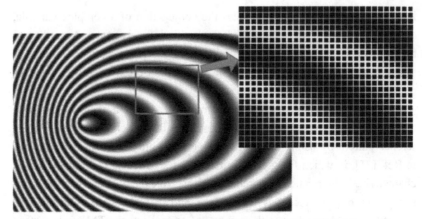

Figure 7.11 The holographic interferograms based on artificial impedance surface.

in which n is the surface refractive index that we discussed in Sections 7.1 and 7.2.

Hence, holographic impedance surface is the holographic interference patterns characterized by using the element of the impedance surface, and the detail is shown in Fig. 7.11. The diagram is full of light and dark stripes, and the grey level of each strip shows the surface impedance of the corresponding element. And the surface impedance is controlled by the geometry of the unit structure of the impedance surface. Equation (7.26) shows the relationship between the surface impedance and the surface refractive index.

In Fig. 7.11, the holographic interference pattern is made up of many quasiperiodic, different-sized surface impedance elements. Different metal patches correspond to different surface impedances. The mathematical expression for the holographic interferograms is

$$Z_s = -i[X + M\mathrm{Re}(\psi_{\mathrm{ref}}^* \psi_{\mathrm{obj}})], \qquad (7.27)$$

where Re means the real part, and X and M are the averaged surface impedance and the modulation depth of the surface impedance, respectively. Here, the modulation depth M denotes the difference between the maximum surface impedance and the average surface impedance. Equation (7.27) can be rewritten as

$$Z_s = -i[\frac{1}{2}X + \frac{1}{2}M\psi_{\mathrm{ref}}^* \psi_{\mathrm{obj}}] - i[\frac{1}{2}X + \frac{1}{2}M\psi_{\mathrm{obj}}^* \psi_{\mathrm{ref}}]. \qquad (7.28)$$

According to the principle of the holographic antenna, if we excite the holographic interference pattern (reference wave) with impedance distribution, as shown in Eq. (7.27), then the resulting representation wave $|\psi_{\mathrm{ref}}|^2 \psi_{\mathrm{obj}}$ will appear; the objective wave can be realized.

From Fig. 7.11, we can roughly judge that the interference is formed by the plane wave and cylindrical wave interference according to the interference pattern; in particular; the selected reference wave (excited source) is expressed as

$$\psi_{\text{ref}} = \exp(ikn\rho), \tag{7.29}$$

in which n is the averaged surface refractive index of the holographic impedance surface, and ρ is the distance between the position of the impedance surface and the center point. As observed from Eq. (7.29), the wavefront phases of the reference wave satisfy the characteristics of the cylindrical wave. The objective wave is selected as

$$\psi_{\text{obj}} = \exp[ikx \sin(\theta)]. \tag{7.30}$$

The wavefront phase meets the characteristics of the plane wave, and Eq. (7.30) shows that the objective wave is the radiation wave and the radiation direction is θ (the angle to the normal of surface). Inserting Eqs. (7.29) and (7.30) into Eq. (7.27), we get

$$Z_s = -i\{X + M\cos[kn\rho - kx \sin(\theta)]\}. \tag{7.31}$$

In Ref. [75], θ is chosen as 60°, and the radiation pattern, or the holographic reproduction, is shown in Fig. 7.12.

From Eq. (7.31), we observe that the surface impedance belongs to the cosine impedance modulation [77]. And the basic function of the modulation is to transform the inspired surface wave into the space radiation wave. The impedance surface for the holographic antenna is based on the cosine or sine modulation, and its role and the basic principle are equivalent to those of leaky wave antennas based on the negative first-order Floquet mode [77].

7.4.4 Tensor Metasurfaces

Recently, an inhomogeneous tensor metasurface was used to manipulate the radiations of two independent beams with different polarizations [78]. The direct holographic method was adopted to modulate the surface impedance matrix on the tensor metasurface. An electric monopole was used to drive the tensor metasurface.

Based on the direct holographic method [75], to produce a desired beam, the tensor surface impedance of a metasurface is required,

$$\overline{\overline{Z}} = -i\begin{pmatrix} X & 0 \\ 0 & X \end{pmatrix} - i\frac{M}{2}\text{Im}(\boldsymbol{E}_{\text{rad}} \otimes \boldsymbol{J}_{\text{surf}}^* - \boldsymbol{J}_{\text{surf}} \otimes \boldsymbol{E}_{\text{rad}}^*), \tag{7.32}$$

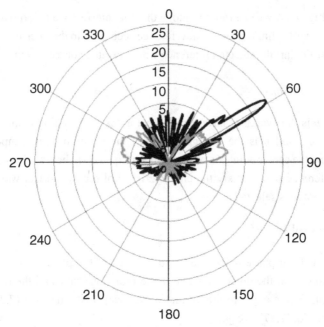

Figure 7.12 Radiation pattern of holographic antenna. The black line means the radiation pattern of the holographic impedance surface excited by a source, and the gray line means the radiation pattern of the same-sized metal plate excited by a source [75].

in which X is the averaged impedance, M is the modulation index, E_{rad} is the expected radiation field, and J_{surf} is the surface current on the metasurface excited by a source. Fig. 7.13(a) shows the illustration of a tensor metasurface, which radiates a directional beam with the left-handed circular polarization (LHCP). To construct a metasurface matching the requested tensor surface impedance, a slotted metal patch is used as the composition element, and its detailed geometry is presented in Fig. 7.13(b), and the distribution of surface-impedance components is shown in Figs. 7.13(c–e). An oblique slot on the metal patch provides the necessary anisotropy so as to form the tensor metasurface.

In Ref. [78], an example of a fully anisotropic metasurface was reported. The azimuth and elevation angles of the directional beam are chosen as $\varphi = 0°$ and $\theta = 45°$, hence the electric field E_{rad} can be written as: $E_{rad} = (i/\sqrt{2}, \ 1, \ -i/\sqrt{2})e^{ik_0 x/\sqrt{2}}$, where k_0 is the wave number in free space. For a point source located at the center of the metasurface, the induced surface current can be described as $J_{surf} = (x, y, 0)e^{ikr}/r$, where k is the wave number on the metasurface, and $r = \sqrt{x^2 + y^2}$. Substituting E_{rad} and J_{surf} into Eq. (7.32), with other parameters chosen as $X = 100$, $M = 0.2$, and $k = 1.1k_0$, the corresponding tensor surface impedance of the metasurface can be calculated.

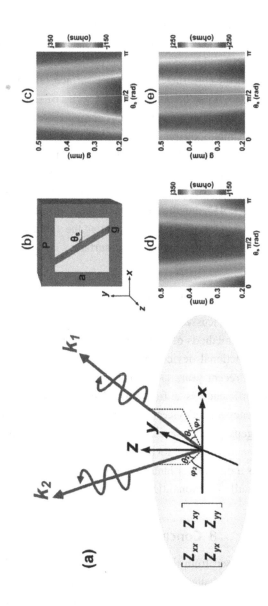

Figure 7.13 (a) Schematic of tensor metasurfaces, in which the subscript "1" represents the left-handed circular polarization of the radiated beam, while the subscript "2" represents the right-handed circular polarization of the radiated beam. (b) Detailed geometries of a metasurface particle. (c) The distribution of Z_{xx}. (d) The distribution of Z_{yy}. (e) The distribution of Z_{xy} [78].

In the design of the tensor metasurface, the substrate is selected as Rogers 3010 with relative permittivity of 10.2 and tangent of loss angle 0.003, and the thickness of the substrate is 1.27 mm. By varying the slot angle θ_s, the patch width a, and the slot width g, the authors established a database of the tensor impedances corresponding to different sets of the geometrical parameters. After establishing the database of tensor surface impedances, practical particles are mapped to the calculated impedance values by using the least squares method. Fig. 7.14(a) shows the constructed tensor metasurface, and the inset is a portion of the metasurface. It can be observed that a series of slotted metallic patches with different geometries constitute a circular-aperture metasurface with a radius of 60 mm. A monopole antenna is placed at the center to excite the metasurface. In Fig. 7.14(b), the grey line illustrates the simulated gain pattern of the LHCP radiation wave, while the black line displays the gain pattern of the right-handed circular polarized (RHCP) wave. The main lobe points to the designed elevation angle 45°, and the LHCP component of the radiation is apparently much larger than the RHCP part, manifesting the validity in design of the polarization using the modulated tensor metasurface.

7.5 Summary

In this section we have discussed the analysis methods, the effective parameters, and the design methods of metasurfaces. Based on the advanced metasurfaces, a lot of functional devices and other applications have been proposed and realized in recent years [79–84]. For transmission-type metasurfaces, transmission efficiency is a fundamental concern. Recently, two high-transmission-efficiency metadevices have been realized by using the carefully designed Huygens' metasurfaces. One is a reconfigurable metalens for which the multiple and complex focal spots can be dynamically controlled [79]. The other is a directional Janus metasurface that can achieve the direction-dependent versatile functionalities [85].

8 Conclusion

In this Element we progressed from the classical Drude–Lorentz model and homogenization theory to the scattering parameters retrieval method, and then to field averaging approach. To derive the effective medium parameters, we also revealed the EM energy densities in the artificial metamaterials. Finally, we presented the effective medium theory, the analysis method, and applications of metasurfaces. The effective medium theory and retrieval methods of effective parameters have been used to design and fabricate many interesting

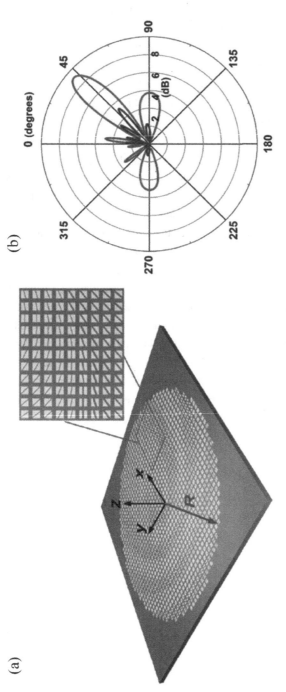

Figure 7.14 Constructed tensor metasurfaces. (a) Tensor metasurface for single beam. Inset is a portion of the metasurface. (b) Gain patterns of LHCP (grey line) and RHCP (black line) components of the radiations [78].

metamaterial devices, such as negative-index lenses [2], invisibility cloaks [6], illusion optical devices [7], metalens-based antennas [86, 87], and so on.

Although we try to offer a systematic and comprehensive review for the effective medium theory of metamaterials and metasurfaces, there are still several unsolved problems in this area. For example, nonlocal or k-dependent homogenization theory should be developed for hyperbolic metamaterials [88]. And how to capture quantum effects in effective medium theory [89]? How to extract effective bulk or surface nonlinear parameters (second- or third-order susceptibilities) for nonlinear responses of metamaterials or metasurfaces [90]? Therefore, there are still some problems to be investigated.

For metamaterials and metasurfaces, the characterization method is not limited to the effective medium theory. In 2014, coding and digital metamaterials were proposed by Cui et al. Analogous to digital circuit technology, different EM responses of meta-atoms in the coding and digital metasurfaces are characterized by discrete digital states, thereby providing functional controls of far fields and near fields by simply changing the digital coding sequences [91]. Owing to programmable and real-time manipulations of the EM waves, the digital coding and programmable metasurfaces have been widely investigated in the last few years, ranging from functional verifications to device designs and system applications [92–96]. From the information perspective, the digital coding concept can be further extended to establish information metasurfaces [97], based on which some simplified-architecture communication and programmable imaging systems can be well realized.

Abbreviations

C–M equation	Clausius–Mossotti equation
ELC	electric-inductor-capacitor
EM	electromagnetic
FDTD	finite-difference time-domain
GSTC	generalized surface transmission condition
LHCP	left-handed circular polarization
PEC	perfect electric conductor
RHCP	right-handed circular polarization
SRR	split-ring resonator
S-parameter	scattering parameter
TE	transverse electric
TM	transverse magnetic
2D	two-dimensional

References

[1] Cui TJ, Tang WX, Yang XM, Mei ZL, and Jiang WX (2016), *Metamaterials: Beyond Crystals, Noncrystals, and Quasicrystals*. CRC Press, Boca Raton, FL.

[2] Shelby RA, Smith DR, and Schultz S (2001), Experimental verification of a negative index of refraction, *Science*, **292**(5514), 77.

[3] Yen TJ, Padilla WJ, Fang N et al. (2004), Terahertz magnetic response from artificial materials, *Science*, **303**(5663), 1494.

[4] Jiang WX, Qiu C-W, Han TC et al. (2013), Broadband all-dielectric magnifying lens for far-field high-resolution imaging, *Adv. Mater.*, **25**(48), 6963–6968.

[5] Pendry JB, Schurig D, and Smith DR (2006), Controlling electromagnetic fields, *Science*, **312**(5781), 1780.

[6] Schruig D, Mock JJ, Justice BJ et al. (2006), Metamaterial electromagnetic cloak at microwave frequencies, A *Science*, **314**(5801), 977.

[7] Jiang WX, Qiu C-W, Han T, Zhang S, and Cui TJ, (2013) Creation of ghost illusions using wave dynamics in metamaterials, *Adv. Funct.*, **23**(32), 4028–4034.

[8] Smith DR and Pendry JB (2006), Homogenization of metamaterials by field averaging, *J. Opt. Soc. Am. B*, **23**(3), 391.

[9] Smith DR, Vier DC, Koschny T, and Soukoulis CM (2005), Electromagnetic parameter retrieval from inhomogeneous metamaterials, *Phys. Rev. E*, **71**(3), 036617.

[10] Lorentz HA (1952), *The Theory of Electrons and Its Applications to the Phenomena of Light and Radiative Heat*, 2nd ed., Dover, Mineola, NY.

[11] Bohren CF and Huffman DR (1998), *Absorption and Scattering of Light by Small Particles*, Wiley, New York.

[12] Rakic AD, Djurisic AB, Elazar JM, and Majewski ML (1998), Optical properties of metallic films for vertical-cavity optoelectronic devices, *Appl. Opt.*, **37**(22), 5271.

[13] Lorentz HA (1880), Ueber die Beziehung zwischen der Fortpflanzungsgeschwindigkeit des Lichtes und der Korperdichte, *Ann. Phys.*, **9**, 641.

[14] Lorenz L (1880), Ueber die Refractionsconstante, *Ann. Phys.*, **11**, 70.

[15] Jackson JD (1998), *Classical Electrodynamics*, 3rd ed., Wiley, New York.

[16] Born M and Wolf E (1999), *Principles of Optics: Electromagnetic Theory of Propagation, Interference and Diffraction of Light*, 7th ed., Cambridge University Press, Cambridge UK.

[17] Kittel C (2005), *Introduction to Solid State Physics*, 8th ed., Wiley, New York.

[18] Maxwell Garnett JC (1904), Colours in metal glasses and in metallic films, *Philos.Trans. R. Soc., London A*, **203**, 385.

[19] Smith GB (1977), Dielectric constants for mixed media, *J. Phys. D*, **10**(4), L39.

[20] Jayannavar AM and Kumar N (1991), Generalization of Bruggeman's unsymmetrical effective-medium theory to a three component composite, *Phys. Rev. B*, **44**(21), 12014.

[21] Merrill WM, Diaz RE, LoRe MM, Squires MC, and Alexopoulos NG (1999), Effective medium theories for artificial materials composed of multiple sizes of spherical inclusions in a host continuum, *IEEE Trans. Antennas Propag.*, **47**(1), 142.

[22] Chettiar UK and Engheta N (2012), Internal homogenization: Effective permittivity of a coated sphere, *Opt. Express*, **20**(21), 22976.

[23] Giovampaola CD and Engheta N (2014), Digital metamaterials, *Nat. Mater.*, **13**(12), 1115.

[24] Ma HF and Cui TJ (2010), Three-dimensional broadband ground-plane cloak made of metamaterials, *Nat. Commun.* **1**, 21.

[25] Jiang WX and Cui TJ (2011), Radar illusion via metamaterials, *Phys. Rev. E*, **83**(2).

[26] Jiang WX, Cui TJ, Yang XM, Ma HF, and Cheng Q (2011), Shrinking an arbitrary object as one desires using metamaterials, *Appl. Phys. Lett.*, **98**(20).

[27] Smith DR, Schultz S, Markos P, and Soukoulis CM (2002), Determination of effective permittivity and permeability of metamaterials from reflection and transmission coefficients, *Phys. Rev. B*, **65**(19), 195104.

[28] Pendry JB, Holden AJ, Stewart WJ, and Youngs I (1996), Extremely low frequency plasmons in metallic mesostructures, *Phys. Rev. Lett.*, **76**(15), 4773.

[29] Huang F, Jiang T, Ran LX, and Chen HS (2004), Experimental confirmation of negative refractive index of a metamaterial composed of Ω-like metallic patterns, *Appl. Phys. Lett.*, **84**(9), 1537–1539.

[30] Ran L, Huangfu J, Chen HS et al. (2005), Experimental study on several left-handed metamaterials, *PIER*, **51**, 249.

[31] Chen HS, Ran LX, Jiang T et al. (2004), Left-handed materials composed of only S-shaped resonators, *Phys. Rev. E*, **70**(5), 057605.

[32] Chen HS, Ran LX, Jiang T et al. (2005), Magnetic properties of S-shaped split-ring resonators, *PIER*, **51**, 231.

[33] Chen HS, Ran LX, Jiang T et al. (2005), Negative refraction of a combined double S-shaped metamaterial, *Appl. Phys. Lett.*, **86**(15), 151909.

[34] Chen X, Grzegorczyk TM, Wu B-I, Pacheco J Jr, and Kong JA (2004), Robust method to retrieve the constitutive effective parameters of metamaterials, *Phys. Rev. E*, **70**(1), 016608.

[35] Li Z, Aydin K, and Ozbay E (2009), Determination of the effective constitutive parameters of bianisotropic metamaterials from reflection and transmission coefficients, *Phys. Rev. E*, **79**(2), 026610.

[36] Hou LL, Chin JY, Yang XM et al. (2008), Advanced parameter retrievals for metamaterial slabs using an inhomogeneous model, *J. Appl. Phys.*, **103**(6), 064904.

[37] J. A. Kong (2002), Electromagnetic wave interaction with stratified negative isotropic media, *Prog. Electromagn. Res. PIER*, **35**, 1.

[38] Smith DR, Schultz S, Markos P, and Soukoulis CM (2002), Determination of effective permittivity and permeability of metamaterials from reflection and transmission coefficients, *Phys. Rev. B*, **65**(19), 195104.

[39] Yee KS (1966), Numerical solution of initial boundary value problems involving Maxwell's equations in isotropic media, *IEEE Trans. Antennas Propag.* **14**(3), 302.

[40] Smith DR and Pendry JB (2006), Homogenization of metamaterials by field averaging, *J. Opt. Soc. Am. B*, **23**(3), 391.

[41] Liu R, Cui TJ, Huang D, Zhao B, and Smith DR (2007), Description and explanation of electromagnetic behaviors in artificial metamaterials based on effective medium theory, *Phys. Rev. E*, **76**(2), 026606.

[42] Kittel C (1996), *Introduction to Solid State Physics*, 7th ed., Wiley, New York.

[43] Xiong XYZ, Jiang LJ, Markel VA, and Tsukerman I (2013), Surface waves in three-dimensional electromagnetic composites and their effect on homogenization, *Opt. Express*, **21**(9), 10412.

[44] Tsukerman I (2011), Effective parameters of metamaterials: a rigorous homogenization theory via Whitney interpolation, *J. Opt. Soc. Am. B*, **28**(3), 577.

[45] Pors A, Tsukerman I, and Bozhevolnyi SI (2011), Effective constitutive parameters of plasmonic metamaterials: homogenization by dual field interpolation, *Phy. Rev. E*, **84**(1), 016609.

[46] Gozhenko VV, Amert AK, and Whites K (2013), Homogenization of periodic metamaterials by field averaging over unit cell boundaries: use and limitations, *New J. Phys.*, **15**, 043030.

[47] Amert AK, Gozhenko VV, and Whites KW (2012), Calculation of effective material parameters by field averaging over lattices with non-negligible unit cell size, *Appl. Phys. A*, **109**(4), 1007.

[48] Smith DR, Schultz S, Markos P, and Soukoulis CM (2002), Determination of effective permittivity and permeability of metamaterials from reflection and transmission coefficients, *Phys. Rev. B*, **65**(19), 195104.

[49] Cui TJ, Smith DR, and Liu RP (2010), *Metamaterials: Theory, Design, and Applications*, Chapters 3 and 4, Springer US, New York.

[50] Landau LD and Lifshitz EM (1984), *Electrodynamics of Continuous Media*, 2nd ed. Pergamon Press, New York.

[51] Cui TJ and Kong JA (2004), Time-domain electromagnetic energy in a frequency-dispersive left-handed medium, *Phys. Rev. B*, **70**(20), 205106.

[52] Ruppin Q (2002), Electromagnetic energy density in a dispersive and absorptive material, *Phys. Lett. A*, **299**(2–3), 309–312.

[53] Boardman AD and Marinov K (2006), Electromagnetic energy in a dispersive metamaterial, *Phys. Rev. B*, **73**(16), 165110.

[54] Luan PG, Wang YT, Zhang S, and Zhang X (2011), Electromagnetic energy density in a single-resonance chiral metamaterial, *Opt. Lett.*, **36**(5), 675.

[55] Luan PG (2009), Power loss and electromagnetic energy density in a dispersive metamaterial medium, *Phys. Rev. E*, **80**(4), 046601.

[56] Loudon R (1970), The propagation of electromagnetic energy through an absorbing dielectric, *J. Phys. A: Gen. Phys.*, **3**(3), 233.

[57] Fung PCW and Young K (1978), Electric energy density in a dissipative medium by circuit analog, *Am. J. Phys.*, **46**(1), 57.

[58] Tretyakov SA (2005), Electromagnetic field energy density in artificial microwave materials with strong dispersion and loss, *Phys. Lett. A*, **343**(1–3), 231–237.

[59] Civelek C and Bechteler TF (2008), Lagrangian formulation of electromagnetic fields in nondispersive medium by means of the extended Euler–Lagrange differential equation, *Int. J. Eng. Sci.*, **46**(12), 1218–1227.

[60] Kong JA (1986), *Electromagnetic Wave Theory*, Wiley, New York.

[61] Kuester EF, Mohamed MA, Piket-May M, and Holloway CL (2003), Averaged transition conditions for electromagnetic fields at a metafilm, *IEEE Trans. Antennas Propag.*, **51**(10), 2641–2651.

[62] Holloway CL, Mohamed MA, Kuester EF, and Dienstfrey A (2005), Reflection and transmission properties of a metafilm: with an application to a controllable surface composed of resonant particles, *IEEE Trans. Antennas Propag.*, **47**(4), 853–865.

[63] Holloway CL, Kuester EF, Gordon JA et al. (2012), An overview of the theory and applications of metasurfaces: the two-dimensional equivalents of metamaterials, *IEEE Trans. Antennas Mag.*, **54**(2), 10–35.

[64] Patel AM and Grbic A (2011), A printed leaky-wave antenna based on a sinusoidally-modulated reactance surface, *IEEE Trans. Antennas Propag.*, **59**(6), 2087–2096.

[65] Patel AM and Grbic A (2011), Modeling and analysis of printed-circuit tensor impedance surfaces, *IEEE Trans. Antennas Propag.*, **59**(1), 2087–2096.

[66] Patel AM and Grbic A (2013), Effective surface impedance of a Printed-Circuit Tensor Impedance Surface (PCTIS), *IEEE Trans. Microwave Theory Tech.*, **61**(4), 1403–1413.

[67] Patel AM and Grbic A (2014), Transformation electromagnetics devices based on Printed-Circuit Tensor Impedance Surfaces, *IEEE Trans. Microwave Theory Tech.*, **62**(5), 1102–1111.

[68] Falcone F, Lopetegi T, Laso MAG et al. (2004), Babinet principle applied to the design of metasurfaces and metamaterials, *Phys. Rev. Lett.*, **93**(19), 197401.

[69] Yu N, Genevet P, Cats MA et al. (2011), Light propagation with phase discontinuities: generalized laws of reflection and refraction, *Science*, **334**(6054), 333.

[70] Pfeiffer C and Grbic A (2013), Metamaterial Huygens' surfaces: tailoring wave fronts with reflectionless sheets, *Phys. Rev. Lett.*, **110**(19), 197401.

[71] Collin RE (1960), *Field Theory of Guided Waves*, McGraw-Hill, New York.

[72] Kang M, Feng TH, Wang HT, and Li J (2012), Wave front engineering from an array of thin aperture antennas, *Opt. Express*, **20**(14), 15882.

[73] Ni X, Kildishev AV, and Shalaev VM (2013), Metasurface holograms for visible light, *Nat. Commun.*, **4**, 2807.

[74] Kock W (1968), Microwave holography, *Microwaves*, **7**(12), 46–54.

[75] Fong BH, Colburn JS, Ottusch JJ, Visher JL, and Sievenpiper DF (2010), Scalar and tensor holographic artificial impedance surfaces, *IEEE Trans. Antennas Propag.*, **58**(10), 3212.

[76] Sievenpiper D, Schaffner JH, Song HJ, Loo RY, and Tangonan G (2003), Two-dimensional beam steering using an electrically tunable impedance surface, *IEEE Trans. Antennas Propag.*, **51**(10), 2713–2722.

[77] Oliner AA and Hessel A (1959), Guided waves on sinusoidally-modulated reactance surfaces, *IEEE Trans. Antennas Propag.*, **7**(5), 201–208.

[78] Wan X, Chen TY, Zhang Q et al. (2016), Manipulations of dual beams with dual polarizations by full-tensor metasurfaces, *Adv. Opt. Mat.*, **4**(10), 1567–1572. doi: https://doi.org/10.1002/adom.201600111.

[79] Chen K, Feng YJ, Monticone F et al. (2017), A reconfigurable active Huygens' metalens, *Adv. Mat.*, **29**(7), 1606422.

[80] Wang SM, Wu PC, Su VC et al. (2017), Broadband achromatic optical metasurface devices, *Nat. Commun.*, **8**,187.

[81] Tittl A, Leitis A, Liu MK et al. (2018), Imaging-based molecular barcoding with pixelated dielectric metasurfaces, *Science*, **360**(6393), 6393.

[82] Yuan YY, Zhang K, Ding XM et al. (2019), Complementary transmissive ultra-thin meta-deflectors for broadband polarization-independent refractions in the microwave region, *Photonics Res.*, **7**(1), 80–88.

[83] Chen MLLN, Jiang LJ, and Sha, WEI (2017) Ultrathin complementary metasurface for orbital angular momentum generation at microwave frequencies, *IEEE Trans. Antennas Propag.*, **66**(1), 396–400.

[84] Xu P, Jiang WX, Wang SY, and Cui TJ (2018), An ultrathin cross-polarization converter with near unity efficiency for transmitted waves, *IEEE Trans. Antennas Propag.*, **66**(8), 4370–4373.

[85] Chen K, Ding G, Hu G, et al. (2020), Directional Janus metasurface, *Adv. Mater.*, **32**(2), 1906352.

[86] Tao Z, Jiang WX, Ma HF, and Cui TJ (2019), High-gain and high-efficiency GRIN metamaterial lens antenna with uniform amplitude and phase distributions on aperture, *IEEE Trans. Antennas Propag.*, **66**(1), 16–22.

[87] Zhang N, Jiang WX, Ma HF, Tang WX, and Cui TJ (2019), Compact high-performance lens antenna based on impedance-matching gradient-index metamaterials, *IEEE Trans. Antennas Propag.*, **67**(2), 1323–1328.

[88] Noginov MA, Barnakov YA, Zhu G et al. (2009), Bulk photonic metamaterial with hyperbolic dispersion, *Appl. Phys. Lett.*, **94**(15), 151105.

[89] Asai H, Savel'ev S, Kawabata S, and Zagoskin AM (2015), Effects of lasing in a one-dimensional quantum metamaterial, *Phys. Rev. B*, **91**(13), 134513.

[90] You JW, Bongu SR, Bao Q, and Panoiu NC (2019), Nonlinear optical properties and applications of 2D materials: theoretical and experimental aspects, *Nanophotonics*, **8**(1), 63.

[91] Cui TJ, Qi MQ, Wan X, Zhao J, and Cheng Q (2014), Coding metamaterials, digital metamaterials and programmable metamaterials, *Light-Science and Applications*, **3**, e218.

[92] Li LL, Cui TJ, Ji W et al. (2017), Electromagnetic reprogrammable coding-metasurface holograms, *Nat. Commun.*, **8**, 197.

[93] Zhang XG, Tang WX et al. (2018), Light-controllable digital coding metasurfaces, *Adv. Sci.*, **5**(11), 1801028.

[94] Zhang XG, Jiang WX, Jiang HL et al. (2020), An optically driven digital metasurface for programming electromagnetic functions, *Nat. Electron.*, **3**(3), 165–171.

[95] Cui TJ, Liu S, Bai GD, and Ma Q (2019), Direct transmission of digital message via programmable coding metasurface, *Research*, **2019**, 2584509.

[96] Cui TJ (2018), Microwave metamaterials, *Natl. Sci. Rev.* **5**(2), 134–136.

[97] Cui TJ, Liu S, and Zhang L (2017), Information metamaterials and metasurfaces, *J. Mater. Chem. C*, **5**(15), 3644–3668.

Acknowledgments

This work was supported in part by the National Key Research and Development Program of China (Grant Nos. 2017YFA0700200 and 2017YFA0700201), in part by the National Natural Science Foundation of China (Grant Nos. 61522106, 61631007, 61571117, 61501112, 61501117, 61401089, 61731010, 61735010, 61722106, 61701107, and 61701108), in part by the Foundation for the Author of National Excellent Doctoral Dissertation of China (201444), and in part by the 111 Project (Grant No. 111-2-05).

Cambridge Elements ᵀ

Emerging Theories and Technologies in Metamaterials

Tie Jun Cui

Southeast University, China

Tie Jun Cui is Cheung-Kong Professor and Chief Professor at Southeast University, China, and a Fellow of the IEEE. He has made significant contributions to the area of effective-medium metamaterials and spoof surface plasmon polaritons at microwave frequencies, both in new-physics verification and engineering applications. He has recently proposed digital coding, field-programmable, and information metamaterials, which extend the concept of metamaterial.

John B. Pendry

Imperial College London

Sir John Pendry is Chair in Theoretical Solid State Physics at Imperial College London, and a Fellow of the Royal Society, the Institute of Physics and the Optical Society of America. Among his many achievements are the proposal of the concepts of an 'invisibility cloak' and the invention of the transformation optics technique for the control of electromagnetic fields.

About the Series

This series systematically covers the theory, characterisation, design and fabrication of metamaterials in areas such as electromagnetics and optics, plasmonics, acoustics and thermal science, nanoelectronics, and nanophotonics, and also showcases the very latest experimental techniques and applications. Presenting cutting-edge research and novel results in a timely, indepth and yet digestible way, it is perfect for graduate students, researchers, and professionals working on metamaterials.

Printed in the United States
by Baker & Taylor Publisher Services